Building
superstructure

General editor: Colin Bassett B.Sc., F.C.I.O.B., F.F.B.

Publisher's note

The contents of this book were originally published in the author's larger four-volume work entitled *Construction Technology* (1973, 1974, 1976 and 1977), (second edition of all four volumes 1987). These chapters have been reproduced in a single volume reference book to meet the well-established need for such a volume.

Note regarding Building Regulations

The reader's attention is drawn to the fact that the Building Regulations, British Standards, Codes of Practice and similar documents are constantly under review and are therefore often revised or amended. It is therefore important that all the Building Regulations, British Standards, Codes of Practice and similar documents quoted in this book are checked by the reader to ensure that the current regulations or recommendations are known and used in practice. Changes in regulations and recommendations often change specific data but the basic principles embodied within them often remain unchanged.

Building superstructure

Second edition

R. Chudley M.C.I.O.B.
Chartered Builder

Illustrated by the author

Longman
Scientific &
Technical

Longman Scientific & Technical,
Longman Group UK Limited,
Longman House, Burnt Mill, Harlow,
Essex CM20 2JE, England
and Associated Companies throughout the world.

© Construction Press 1982
This edition © Longman Group UK Limited 1988

First published 1982 by Construction Press
Reprinted 1985
Second edition 1989
Fourth impression 1993

British Library Cataloguing in Publication Data
Chudley, R. (Roy)
   Building and superstructure. — 2nd ed.
   1. Buildings. Structural components
   I. Title
   690

ISBN 0-582-01969-9

Printed in Malaysia by VP

Contents

Preface

The compilation of this book has been carried out with the main objective of providing a reference book for the practical and practising builder. It may also be a useful reference work for building students of all levels engaged in assignment and project work. The contents have been arranged to cover the basic structural elements of buildings above ground level together with temporary works such as shoring and scaffolding.

The contents of this book are based on typical construction technology concepts. Technology can be defined as the science of mechanical and industrial arts, as contrasted with fine arts. Similarly science can be defined as an ordered arrangement of facts under classes or headings, theoretical knowledge as distinguished from practical knowledge and knowledge of principles and rules. The information contained within this book is therefore based on the principles and techniques of construction and not on actual case studies of works in progress, however, if experience and practical knowledge are added to the technological concepts the result should provide buildings and construction techniques of a high and therefore acceptable standard.

The reader's attention is particularly drawn to the note at the beginning of the text regarding current Building Regulations.

R. Chudley
Guildford 1982

Acknowledgements

We are grateful to the following for permission to reproduce copyright material:

British Standards Institution for reference to British Standards Codes of Practice; Building Research Station for extracts from *Building Research Station Digests*; Her Majesty's Stationery Office for extracts from Acts, Regulations and Statutory Instruments.

Part I

Temporary works

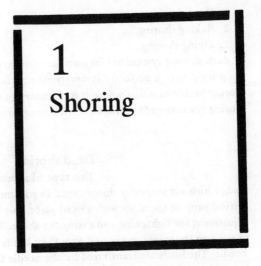

1
Shoring

All forms of shoring are temporary supports applied to a building or structure to comply with the requirements of Regulation 50 of the Construction (General Provisions) Regulations 1961. This regulation requires that all practical precautions shall be taken to avoid danger to any person from collapse of structure. Common situations where shoring may be required are:

1. To give support to walls which are dangerous or are likely to become unstable due to subsidence, bulging or leaning.
2. To avoid failure of sound walls caused by the removal of subjacent support such as where a basement is being constructed near to a sound wall.
3. During demolition works to give support to an adjacent building or structure.
4. To support the upper part of wall during formation of a large opening in the lower section of the wall.
5. To give support to a floor or roof to enable a support wall to be removed and replaced by a beam.

Structural softwood is the usual material used for shoring members; its strength to weight ratio compares favourably with that of structural steel and its adaptability is superior to steel. Shoring arrangements can also be formed by coupling together groups of scaffold tubulars.

SHORING SYSTEMS
There are three basic shoring systems, namely:

1. Dead shoring.
2. Raking shoring.
3. Flying shoring.

Each shoring system has its own function to perform and is based upon the principles of a perfectly symmetrical situation. In practice many shoring problems occur where it is necessary to use combinations of shoring systems and/or unsymmetric arrangements (see Fig. I.1).

Dead shoring

This type of shoring is used to support dead loads which act vertically downwards. In its simplest form it consists of a vertical prop or shore leg with a head plate, sole plate and some means of adjustment for tightening and easing the shore. The usual arrangement is to use two shore legs connected over their heads by a horizontal beam or needle. The loads are transferred by the needle to the shore legs and hence down to a solid bearing surface. It may be necessary to remove pavings and cut holes in suspended timber floors to reach a suitable bearing surface; if a basement is encountered a third horizontal member called a transom will be necessary since it is impracticable to manhandle a shore leg through two storeys. A typical example of this situation is shown in Fig. I.2.

The sequence of operations necessary for a successful dead shoring arrangement can be enumerated thus:

1. Carry out a thorough site investigation to determine:
   (a) number of shores required by ascertaining possible loadings and window positions;
   (b) bearing capacity of soil and floors;
   (c) location of underground services which may have to be avoided or bridged;
2. Fix ceiling struts between suitable head and sole plates to relieve the wall of floor and roof loads. The struts should be positioned as close to the wall as practicable.
3. Strut all window openings within the vicinity of the shores to prevent movement or distortion of the opening. The usual method is to place timber plates against the external reveals and strut between them; in some cases it may be necessary to remove the window frame to provide sufficient bearing surface for the plates.
4. Cut holes through the wall slightly larger in size than the needles.
5. Cut holes through ceilings and floors for the shore legs.
6. Position and level sleepers on a firm base, removing pavings if necessary.

4

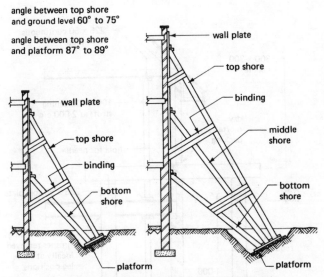

angle between top shore
and ground level 60° to 75°

angle between top shore
and platform 87° to 89°

wall plate

top shore

binding

middle
shore

bottom
shore

wall plate

top shore

binding

bottom
shore

platform

platform

Typical raking shore arrangements (see also Fig. I.11)

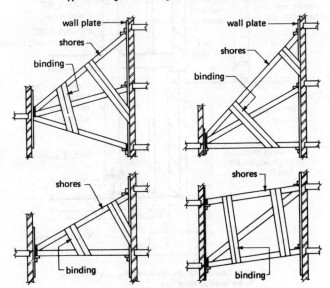

wall plate

shores

binding

wall plate

shores

binding

shores

binding

shores

binding

Unsymmetrical flying shore arrangements

Fig. I.1 Shoring arrangements

5

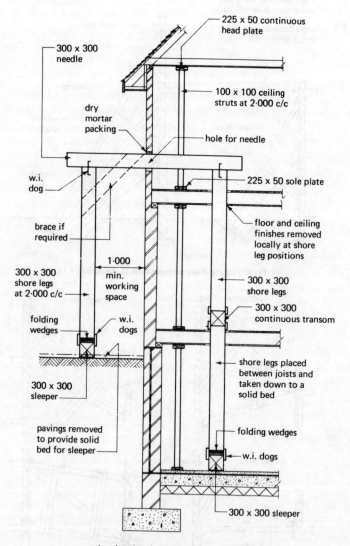

225 x 50 continuous head plate

300 x 300 needle

100 x 100 ceiling struts at 2·000 c/c

dry mortar packing

hole for needle

w.i. dog

225 x 50 sole plate

brace if required

floor and ceiling finishes removed locally at shore leg positions

1·000 min. working space

300 x 300 shore legs at 2·000 c/c

300 x 300 shore legs

folding wedges

w.i. dogs

300 x 300 continuous transom

300 x 300 sleeper

shore legs placed between joists and taken down to a solid bed

pavings removed to provide solid bed for sleeper

folding wedges

w.i. dogs

300 x 300 sleeper

cross bracing, longitudinal bracing and hoardings to be fixed as necessary

Fig. I.2 Dead shoring

7. Erect, wedge and secure shoring arrangements.

Upon completion of the builder's work it is advisable to leave the shoring in position for at least seven days before easing the supports to ensure the new work has gained sufficient strength to be self supporting.

Raking shores

This shoring arrangement transfers the floor and wall loads to the ground by means of sloping struts or rakers. It is very important that the rakers are positioned correctly so that they are capable of receiving maximum wall and floor loads. The centre line of the raker should intersect with the centre lines of the wall or floor bearing; common situations are detailed in Fig. I.3. One raker for each floor is required and ideally should be at an angle of between 40° and 70° with the horizontal; therefore the number of rakers which can be used is generally limited to three. A four-storey building can be shored by this method if an extra member, called a rider, is added (see Fig. I.4).

The operational sequence for erecting raking shoring can be enumerated thus:

1. Carry out site investigation as described for dead shoring.
2. Mark out and cut mortices and housings in wall plate.
3. Set out and cut holes for needles in external wall.
4. Excavate to a firm bearing subsoil and lay grillage platform and sole plate.
5. Cut and erect rakers commencing with the bottom shore. A notch is cut in the heel so that a crow bar can be used to lever the raker down the sole plate and thus tighten the shore. The angle between sole plate and shores should be at its maximum about 89° to ensure that the tangent point is never reached and not so acute that levering is impracticable.
6. Fix cleats, distance blocks, binding and if necessary cross bracing over the backs of the shores.

Flying shores

These shores fulfil the same functions as a raking shore but have the advantage of providing a clear working space under the shoring. They can be used between any parallel wall surfaces providing the span is not in excess of 12.000 m when the arrangement would become uneconomic. Short spans up to 9.000 m usually have a single horizontal member whereas the larger spans require two horizontal shores to keep the section sizes within the timber range commercially available (see Fig. I.5 and Fig. I.6).

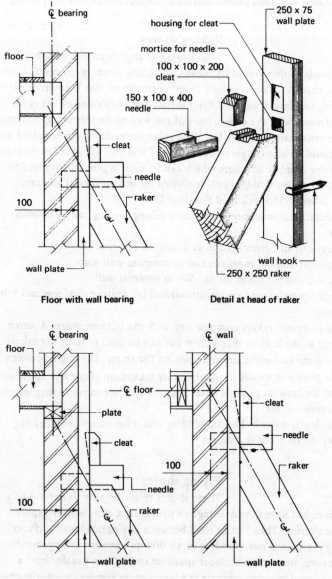

Labels in figure:

₵ bearing

floor

100

wall plate

cleat

needle

raker

Floor with wall bearing

housing for cleat

mortice for needle

100 × 100 × 200
cleat

150 × 100 × 400
needle

250 × 75
wall plate

wall hook

250 × 250 raker

Detail at head of raker

₵ bearing

floor

plate

cleat

needle

raker

100

wall plate

Floor with plate bearing

₵ floor

wall

cleat

needle

raker

100

wall plate

Floor parallel to wall

Fig. I.3 Raking shore intersections

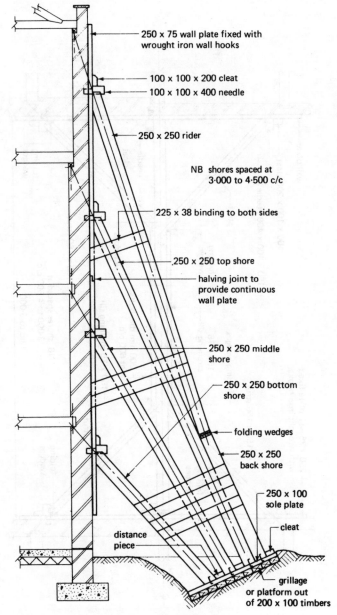

250 x 75 wall plate fixed with wrought iron wall hooks

100 x 100 x 200 cleat
100 x 100 x 400 needle

250 x 250 rider

NB shores spaced at 3·000 to 4·500 c/c

225 x 38 binding to both sides

250 x 250 top shore

halving joint to provide continuous wall plate

250 x 250 middle shore

250 x 250 bottom shore

folding wedges

250 x 250 back shore

250 x 100 sole plate

cleat

distance piece

grillage or platform out of 200 x 100 timbers

Fig. I.4 Typical multiple raking shore

9

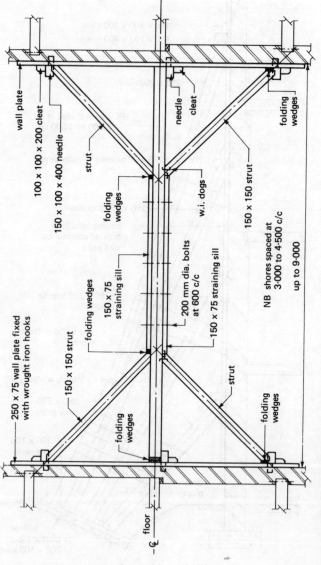

wall plate

100 x 100 x 200 cleat

150 x 100 x 400 needle

strut

folding wedges

150 x 75 straining sill

folding wedges

250 x 75 wall plate fixed with wrought iron hooks

150 x 150 strut

folding wedges

needle

cleat

folding wedges

150 x 150 strut

w.i. dogs

200 mm dia. bolts at 600 c/c

150 x 75 straining sill

NB shores spaced at 3·000 to 4·500 c/c up to 9·000

150 x 150 strut

strut

folding wedges

floor

₵

Fig. I.5 Typical single flying shore

10

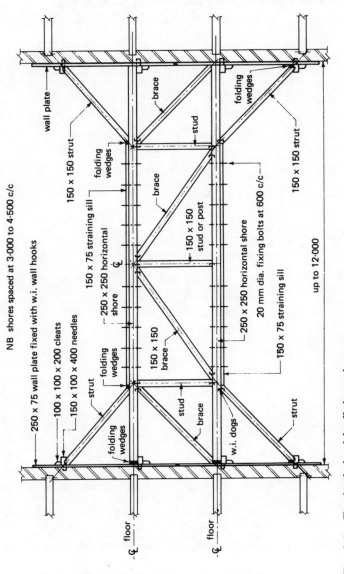

NB shores spaced at 3·000 to 4·500 c/c

250 x 75 wall plate fixed with w.i. wall hooks

100 x 100 x 200 cleats

150 x 100 x 400 needles

strut

folding wedges

150 x 150 brace

stud

brace

w.i. dogs

strut

wall plate

brace

folding wedges

stud

150 x 150 strut

folding wedges

150 x 75 straining sill

250 x 250 horizontal shore

℄

150 x 150 stud or post

20 mm dia. fixing bolts at 600 c/c

250 x 250 horizontal shore

150 x 75 straining sill

150 x 150 strut

up to 12·000

℄ floor

℄ floor

Fig. I.6 Typical double flying shore

11

It is possible with all forms of shoring to build up the principal members from smaller sections by using bolts and timber connectors, ensuring all butt joints are well staggered to give adequate rigidity. This in effect is a crude form of laminated timber construction.

The site operations for the setting out and erection of a flying shoring system are similar to those enumerated for raking shoring.

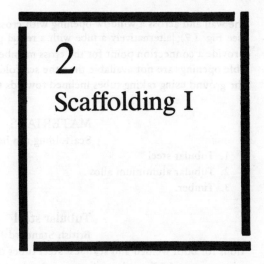

2
Scaffolding I

A scaffold is a temporary structure from which persons can gain access to a place of work in order to carry out building operations, it includes any working platforms, ladders and guard rails. Basically there are two forms of scaffolding:

1. Putlog scaffolds.
2. Independent scaffolds.

PUTLOG SCAFFOLDS

This form of scaffolding consists of a single row of uprights or standards set away from the wall at a distance which will accommodate the required width of the working platform. The standards are joined together with horizontal members called ledgers and are tied to the building with cross members called putlogs. The scaffold is erected as the building rises and is mostly used for buildings of traditional brick construction (see Fig. I.7).

INDEPENDENT SCAFFOLDS

An independent scaffold has two rows of standards which are tied by cross members called transoms. This form of scaffold does not rely upon the building for support and is therefore suitable for use in conjunction with framed structures (see Fig. I.8).

Every scaffold should be securely tied to the building at intervals of approximately 3.600 m vertically and 6.000 m horizontally. This can be achieved by using a horizontal tube called a bridle bearing on the inside of

13

the wall and across a window opening with cross members connected to it (see Fig. I.7); |alternatively a tube with a reveal pin in the opening can provide a connection point for the cross members (see Fig. I.8). If suitable openings are not available then the scaffold should be strutted from the ground using raking tubes inclined towards the building.

MATERIALS
Scaffolding can be of:

1. Tubular steel.
2. Tubular aluminium alloy.
3. Timber.

Tubular steel
British Standard 1139 gives recommendations for both welded and seamless steel tubes of 48 mm outside diameter with a nominal 38 mm bore diameter. Steel tubes can be obtained galvanised (to guard against corrosion); ungalvanised tubes will require special care such as painting, varnishing or an oil bath after use. Steel tubes are nearly three times heavier than comparable aluminium alloy tubes but are far stronger and since their deflection is approximately one third of aluminium alloy tubes, longer spans can be used.

Aluminium alloy
Seamless tubes of aluminium alloy with a 48 mm outside diameter are specified in BS 1139 for metal scaffolding. No protective treatment is required unless they are to be used in contact with materials such as damp lime, wet cement and sea water, which can cause corrosion of the aluminium alloy tubes. A suitable protective treatment would be to coat the tubes with bitumastic paint before use.

Timber
The use of timber as a temporary structure in the form of a scaffold is now rarely encountered in this country, although it is still used extensively in other countries. The timber used is fir of structural quality in either putlog or independent format, the members being lashed together with wire or rope instead of the coupling fittings used with metal scaffolds.

Scaffold boards
These are usually boards of softwood timber complying with the recommendations of BS 2482 used to form the working platform at the required level. They should be formed out of specified

softwoods of 225 × 38 section and not exceeding 4.800 m in length. To prevent the ends from splitting they should be end bound with not less than 25 mm wide × 0.9 mm galvanised hoop iron extending at least 150 mm along each edge and fixed with a minimum of two fixings to each end. The strength of the boards should be such that they can support a uniformity distributed load of 6.7 kN/m² when supported at 1.200 m centres.

Scaffold fittings

Fittings of either steel or aluminium alloy are covered by the same British Standard as quoted above for the tubes. They can usually be used in conjunction with either tubular metal unless specified differently by the manufacturer. The major fittings used in metal scaffolding are:

Double coupler: the only real load bearing fitting used in scaffolding and is used to join ledgers to standards.

Swivel coupler: composed of two single couplers riveted together so that it is possible to rotate them and use them for connecting two scaffold tubes at any angle.

Putlog coupler: used solely for fixing putlogs or transoms to the horizontal ledgers.

Base plate: a square plate with a central locating spigot used to distribute the load from the foot of a standard on to a sole plate or firm ground. Base plates can also be obtained with a threaded spigot and nut for use on sloping sites to make up variations in levels.

Split joint pin: a connection fitting used to joint scaffold tubes end to end. A centre bolt expands the two segments which grip on the bore of the tubes.

Reveal pin: fits into the end of a tube to form an adjustable strut.

Putlog end: a flat plate which fits on the end of a scaffold tube to convert it into a putlog.

Typical examples of the above fittings are shown in Fig. I.9.

THE CONSTRUCTION (WORKING PLACES) REGULATIONS 1966

This statutory instrument is designed to ensure that suitable and sufficient safe access to and egress from every

15

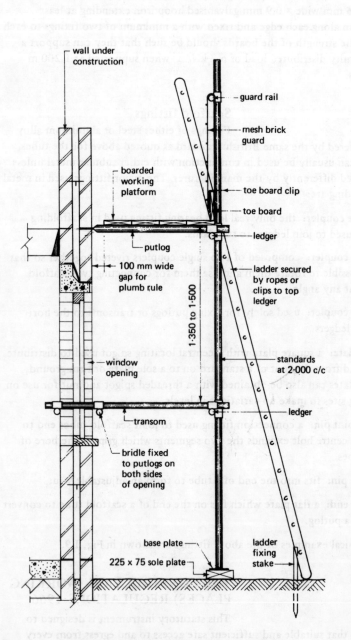

wall under construction

guard rail

mesh brick guard

boarded working platform

toe board clip

toe board

ledger

putlog

ladder secured by ropes or clips to top ledger

100 mm wide gap for plumb rule

1·350 to 1·500

window opening

standards at 2·000 c/c

ledger

transom

bridle fixed to putlogs on both sides of opening

base plate

225 x 75 sole plate

ladder fixing stake

Fig. I.7 Typical tubular steel putlog scaffold

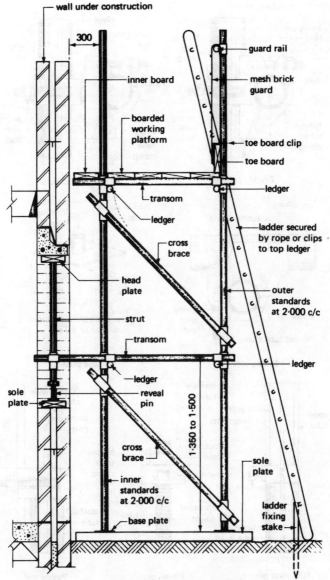

N.B. Not more than 50% of ties should be reveal ties

Fig. I.8 Typical tubular steel independent scaffold

17

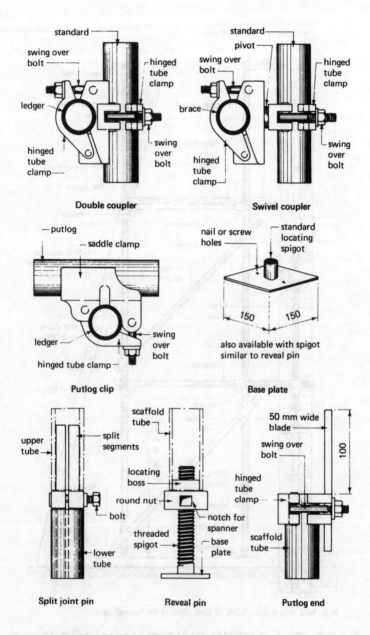

Fig. I.9 Typical steel scaffold fittings

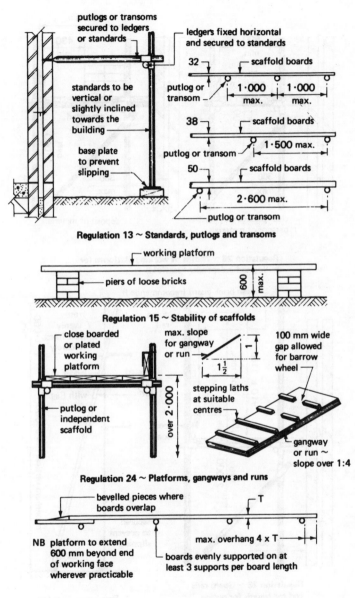

putlogs or transoms
secured to ledgers
or standards

ledgers fixed horizontal
and secured to standards

standards to be
vertical or
slightly inclined
towards the
building

base plate
to prevent
slipping

32 — scaffold boards

putlog or transom — 1·000 max. 1·000 max.

38 — scaffold boards

putlog or transom — 1·500 max.

50 — scaffold boards

2·600 max.

putlog or transom

Regulation 13 ~ Standards, putlogs and transoms

working platform

piers of loose bricks

600 max.

Regulation 15 ~ Stability of scaffolds

close boarded
or plated
working
platform

max. slope
for gangway
or run →

1
1½

100 mm wide
gap allowed
for barrow
wheel

over 2·000

putlog or
independent
scaffold

stepping laths
at suitable
centres

gangway
or run ~
slope over 1:4

Regulation 24 ~ Platforms, gangways and runs

bevelled pieces where
boards overlap

T

NB platform to extend
600 mm beyond end
of working face
wherever practicable

max. overhang 4 x T

boards evenly supported on at
least 3 supports per board length

Regulation 25 ~ Boards in working platforms

Fig. I.10 Scaffolds and Construction Regulations — 1

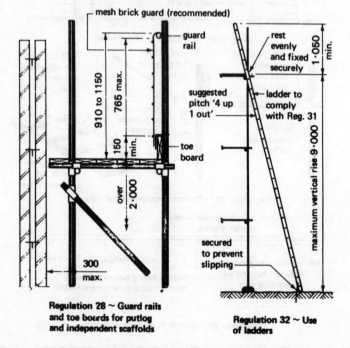

Regulation 26 ~ Widths of working platforms for putlog and independent scaffolds

Regulation 28 ~ Guard rails and toe boards for putlog and independent scaffolds

Regulation 32 ~ Use of ladders

Fig. I.11   Scaffolds and Construction Regulations — 2

place at which any person at any time works is provided and properly maintained. Scaffolds and ladders are covered by this document which sets out the minimum requirements for materials, maintenance, inspection and construction of these working places. The main constructional requirements of these regulations are illustrated in Figs. I.10 and I.11. The metric dimensions shown in the figures are those quoted in the Statutory Instrument entitled *The Construction (Metrication) Regulations 1984* as being acceptable metric equivalents of the imperial dimensions given in the actual regulations.

Regulation 22 sets out the requirements for the inspection of scaffolds, by a competent person, which are:
1. Within the preceding seven days.
2. After adverse weather conditions which may have affected the scaffold's strength or stability.

A record of all such inspections must be kept in accordance with Regulation 39 and must give the following information:
1. Location and description of scaffold.
2. Date of inspection.
3. Result of inspection, stating the condition of the scaffold.
4. Signature of person who made the inspection.

The importance of providing a safe and reliable scaffold from which to carry out building operations cannot be over emphasised, since badly constructed and non-maintained scaffolds are a large contributory factor to the high accident rate which prevails in the building industry today.

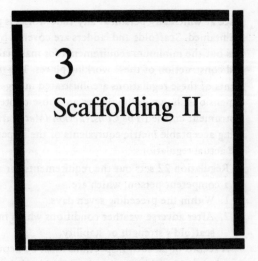

3
Scaffolding II

A scaffold is a temporary frame usually constructed from steel or aluminium alloy tubes clipped or coupled together to provide a means of access to high level working areas as well as providing a safe platform from which to work. The two basic forms of scaffolding, namely the putlog scaffold with its single row of uprights or standards set outside the perimeter of the building and partly supported by the structure and independent scaffolds which have two rows of standards should have been covered in the second year of study (see Chapter 2).

It is therefore only necessary to consider in this text the special scaffolds such as slung, suspended, truss-out and gantry scaffolds as well as the easy-to-erect system scaffolds. It cannot be over-emphasised that all scaffolds must fully comply with the minimum requirements set out in the Construction (Working Places) Regulations 1966 to which reference must be made (see Chapter 2).

Slung scaffolds: these are scaffolds which are suspended by means of wire ropes or chains and are not provided with a means of being raised or lowered by a lifting appliance. Their main use is for gaining access to high ceilings or the underside of high roofs. A secure anchorage must be provided for the suspension ropes and this can usually be achieved by using the structural members of the roof over the proposed working area. Any member selected to provide the anchorage point must be inspected to assess its adequacy. At least six evenly spaced suspension wire ropes or chains should be used and these must be adequately secured at both ends.

The working platform is constructed in a similar manner to conventional scaffolds, consisting of ledgers, transoms, and timber scaffold boards with the necessary guard rails and toe boards. Working platforms in excess of 2.400 x 2.400 plan size should be checked to ensure that the supporting tubular components are not being overstressed.

Truss-out scaffolds: these are a form of an independent tied scaffold which rely entirely on the building for support and are used where it is impossible or undesirable to erect a conventional scaffold from ground level. The supporting scaffolding structure which projects from the face of the building is known as the truss-out. Anchorage is provided by adjustable struts fixed internally between the floor and ceiling from which projects the cantilever tubes. Except for securing rakers only right-angle couplers should be used. The general format for the remainder of the scaffold is as used for conventional independent scaffolds — see Fig. I.12.

Suspended scaffolds: these consist of a working platform suspended from supports such as outriggers which cantilever over the upper edge of a building and in this form are a temporary means of access to the face of a building for the purposes of cleaning and light maintenance work. Many new tall structures have suspension tracks incorporated in the fascia or upper edge beam, or a cradle suspension track is fixed to the upper surface of the flat roof on which is supported a manual or power trolley with retractable davit arms for supporting the suspended working platform or cradle. All forms of suspended cradles must conform with the minimum requirements set out in the Construction (Working Places) Regulations 1966 with regard to platform boards, guard rails and toe boards. Cradles may be single units or grouped together to form a continuous working platform; if grouped together they are connected to one another at their abutment ends with hinges which form a gap not exceeding 25 mm wide. Figure I.13 shows typical suspended scaffold details.

Mobile tower scaffolds: these are used mainly by painters and maintenance staff to gain access to ceilings where it is advantageous to have a working platform which can be readily moved to a new position. The scaffold is basically a square tower constructed from scaffold tubes mounted on wheels fitted with brakes. Access is gained by means of a vertical ladder securely fixed to one side of the tower. A working platform complying with all the relevant regulations with a plan size of not less than 1.200 x 1.200 should be provided. To ensure complete stability the height of the tower should not exceed three-and-a-half times its least lateral dimension for internal work and three times its least lateral dimension for external work with a maximum height of 10.000 unless tied to the structure — see Fig. I.14 for typical details.

23

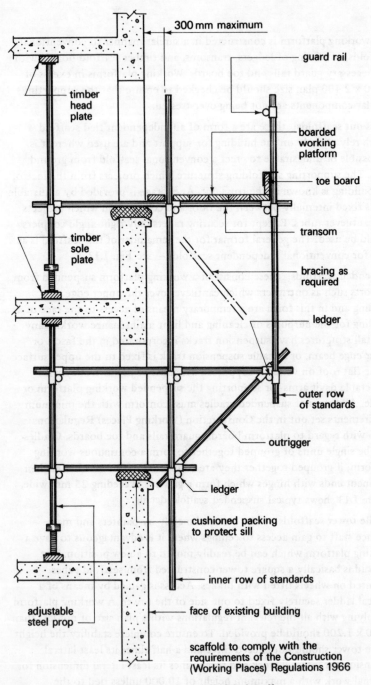

300 mm maximum

guard rail

timber head plate

boarded working platform

toe board

transom

timber sole plate

bracing as required

ledger

outer row of standards

outrigger

ledger

cushioned packing to protect sill

inner row of standards

adjustable steel prop

face of existing building

scaffold to comply with the requirements of the Construction (Working Places) Regulations 1966

Fig. I.12 Typical truss-out scaffold details

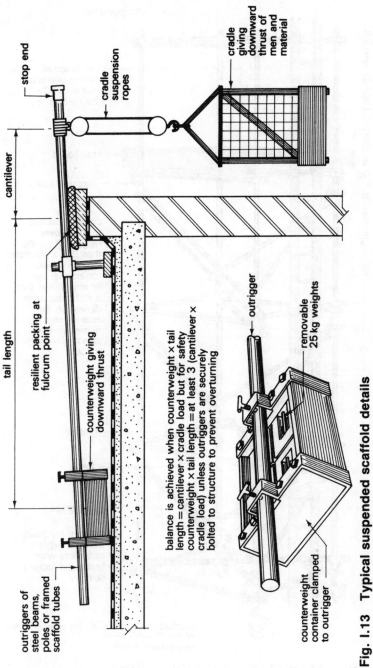

stop end

cradle
suspension
ropes

cradle
giving
downward
thrust of
men and
material

cantilever

resilient packing at
fulcrum point

counterweight giving
downward thrust

tail length

balance is achieved when counterweight × tail
length = cantilever × cradle load but for safety
counterweight × tail length = at least 3 (cantilever ×
cradle load) unless outriggers are securely
bolted to structure to prevent overturning

outriggers of
steel beams,
poles or framed
scaffold tubes

outrigger

removable
25 kg weights

counterweight
container clamped
to outrigger

Fig. I.13 Typical suspended scaffold details

25

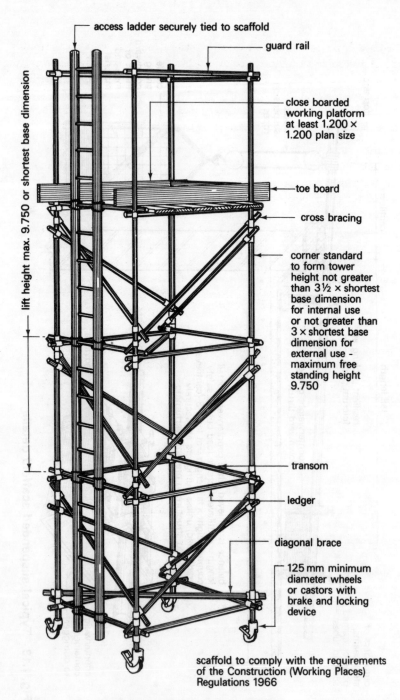

access ladder securely tied to scaffold

guard rail

close boarded working platform at least 1.200 × 1.200 plan size

toe board

cross bracing

corner standard to form tower height not greater than 3½ × shortest base dimension for internal use or not greater than 3 × shortest base dimension for external use – maximum free standing height 9.750

transom

ledger

diagonal brace

125 mm minimum diameter wheels or castors with brake and locking device

lift height max. 9.750 or shortest base dimension

scaffold to comply with the requirements of the Construction (Working Places) Regulations 1966

Fig. I.14 Typical mobile tower scaffold

Birdcage scaffolds: these are used to provide a complete working platform at high level over a large area and consist basically of a two directional arrangement of standards, ledgers and transoms to support a close boarded working platform at the required height. To ensure adequate stability standards should be placed at not more than 2.400 centres in both directions and the whole arrangement adequately braced.

Gantries: these are forms of scaffolding used primarily as elevated loading and unloading platforms over a public footpath where the structure under construction or repair is immediately adjacent to the footpath. As in the case of hoardings local authority permission is necessary and their specific requirements such as pedestrian gangways, lighting and dimensional restrictions must be fully met. It may also be necessary to comply with police requirements as to when loading and unloading can take place. The gantry platform can also serve as a storage and accommodation area as well as providing the staging from which a conventional independent scaffold to provide access to the face of the building can be erected. Gantry scaffolds can be constructed from standard structural steel components as shown in Fig. I.15 or from a system scaffold as shown in Fig. I.16.

System scaffolds: these scaffolds are based upon the traditional independent steel tube scaffold but instead of being connected together with a series of loose couplers and clips they usually have integral interlocking connections. They are easy to erect, adaptable and generally they can be assembled and dismantled by semi-skilled labour. The design of these systems is such that the correct position of handrails, lift heights and all other aspects of the Construction (Working Places) Regulations 1966 are automatically met. Another advantage found in most of these system scaffolds is the elimination of internal cross-bracing, giving a clear walk through space at all levels; façade-bracing however may still be required. Fig. I.17 shows details of a typical system scaffold and like the illustrations chosen for items of builders' mechanical plant is only intended to be representative of the many system scaffolds available.

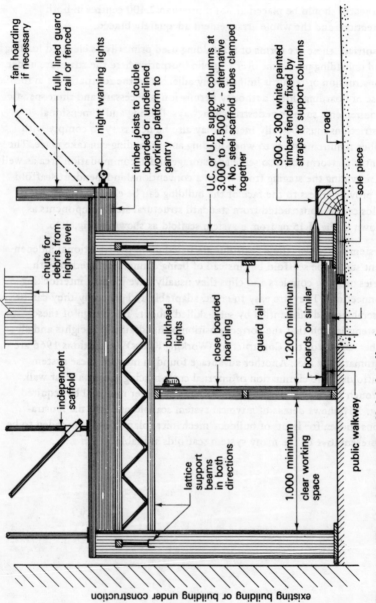

fan hoarding if necessary

fully lined guard rail or fenced

night warning lights

timber joists to double boarded or underlined working platform to gantry

U.C. or U.B. support columns at 3.000 to 4.500 % - alternative 4 No. steel scaffold tubes clamped together

300 × 300 white painted timber fender fixed by straps to support columns

road

sole piece

chute for debris from higher level

independent scaffold

bulkhead lights

close boarded hoarding

guard rail

1.200 minimum

boards or similar

public walkway

lattice support beams in both directions

1.000 minimum clear working space

existing building or building under construction

Fig. I.15 Typical gantry scaffold details 1

28

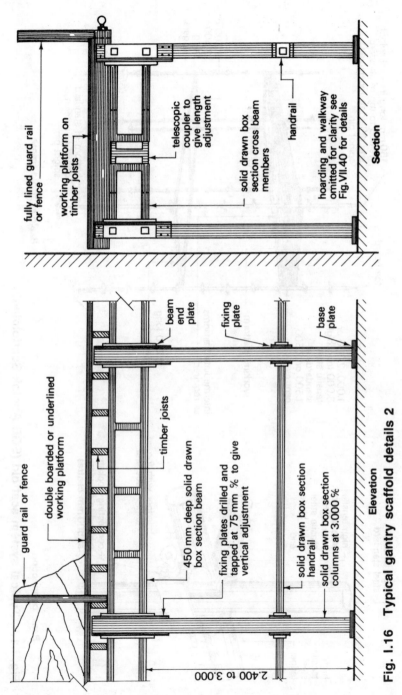

fully lined guard rail or fence

working platform on timber joists

telescopic coupler to give length adjustment

solid drawn box section cross beam members

handrail

hoarding and walkway omitted for clarity see Fig.VII.40 for details

Section

guard rail or fence

double boarded or underlined working platform

beam end plate

fixing plate

base plate

timber joists

450 mm deep solid drawn box section beam

fixing plates drilled and tapped at 75 mm % to give vertical adjustment

solid drawn box section handrail

solid drawn box section columns at 3.000 %

2.400 to 3.000

Elevation

Fig. I.16   Typical gantry scaffold details 2

29

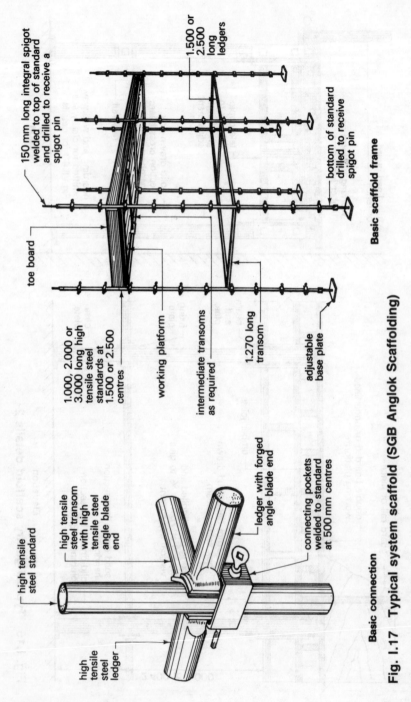

150 mm long integral spigot welded to top of standard and drilled to receive a spigot pin

1.500 or 2.500 long ledgers

bottom of standard drilled to receive spigot pin

Basic scaffold frame

toe board

1.000, 2.000 or 3.000 long high tensile steel standards at 1.500 or 2.500 centres

working platform

intermediate transoms as required

1.270 long transom

adjustable base plate

Basic connection

high tensile steel standard

high tensile steel transom with high tensile steel angle blade end

ledger with forged angle blade end

connecting pockets welded to standard at 500 mm centres

high tensile steel ledger

Fig. I.17 Typical system scaffold (SGB Anglok Scaffolding)

30

Part II

Basic

superstructure

4
Stonework, brickwork and blockwork

STONEWORK

The natural formation of stones or rocks is a very long process which commenced when the earth was composed only of gases. These gases eventually began to liquefy forming the land and the sea, the land being only a thin crust or mantle to the still molten core.

Igneous rocks originated from the molten state; plutonic rocks are those in the lower part of the earth's mantle; whereas hypabyssal rocks solidified rapidly near the upper surface of the crust.

Building stones

Stones used in building can be divided into three classes as follows:

1. Igneous.
2. Sedimentary.
3. Metamorphic.

IGNEOUS STONES

These stones originate from volcanic action being formed by the crystallisation of molten rock matter derived from deep in the earth's crust. It is the proportions of these crystals which give the stones their colour and characteristics. Granites are typical of this class of stone being hard, durable and capable of a fine polished finish. Granites are mainly composed of quartz, felspar and mica.

SEDIMENTARY STONES

These stones are largely composed of material derived from the breakdown and erosion of existing rocks deposited in layers under the waters, which at that time covered much of the earth's surface. Being deposited in this manner their section is stratified and shows to a lesser or greater degree the layers as deposited. Some of these layers are only visible if viewed under a microscope. Under the microscope it will be seen that all the particles lie in one direction indicating the flow of the current in the water. Sandstones and limestones are typical examples of sedimentary stones.

Sandstones are stratified sedimentary rocks produced by the eroded and disintegrated rocks, like granite, being carried away and deposited by water in layers. The brown and yellow tints in sandstones are due to the presence of oxides of iron.

Limestones may be organically formed by the deposit of tiny shells and calcareous (containing limestone) skeletons in the seas and rivers or it may be formed chemically by deposits of lime in ringed layers. Limestones vary considerably from a heavy crystalline form to a friable material such as chalk.

METAMORPHIC STONES

These are stones which have altered and may have been originally igneous or sedimentary rocks, but have since been changed by geological processes such as pressure, movement, heat and chemical reaction due to the infiltration of fluids. Typical examples of this type of stone are marbles and slates.

Marbles are metamorphic limestones, their original structure having been changed by pressure. Marbles being capable of taking a high polish are used mainly for decorative work.

Slate is a metamorphic clay having been subjected to great pressure and heat; being derived from a sedimentary layer it can be easily split into thin members.

Stones are obtained from quarries by blasting and wedging the blocks away from the solid mass. They are partly worked in the quarry and then sent to store yards where they can be sawn, cut, moulded, dressed and polished to the customer's requirements.

Today, natural stones are sometimes used for facing prestige buildings, constructing boundary or similar walls and in those areas where natural

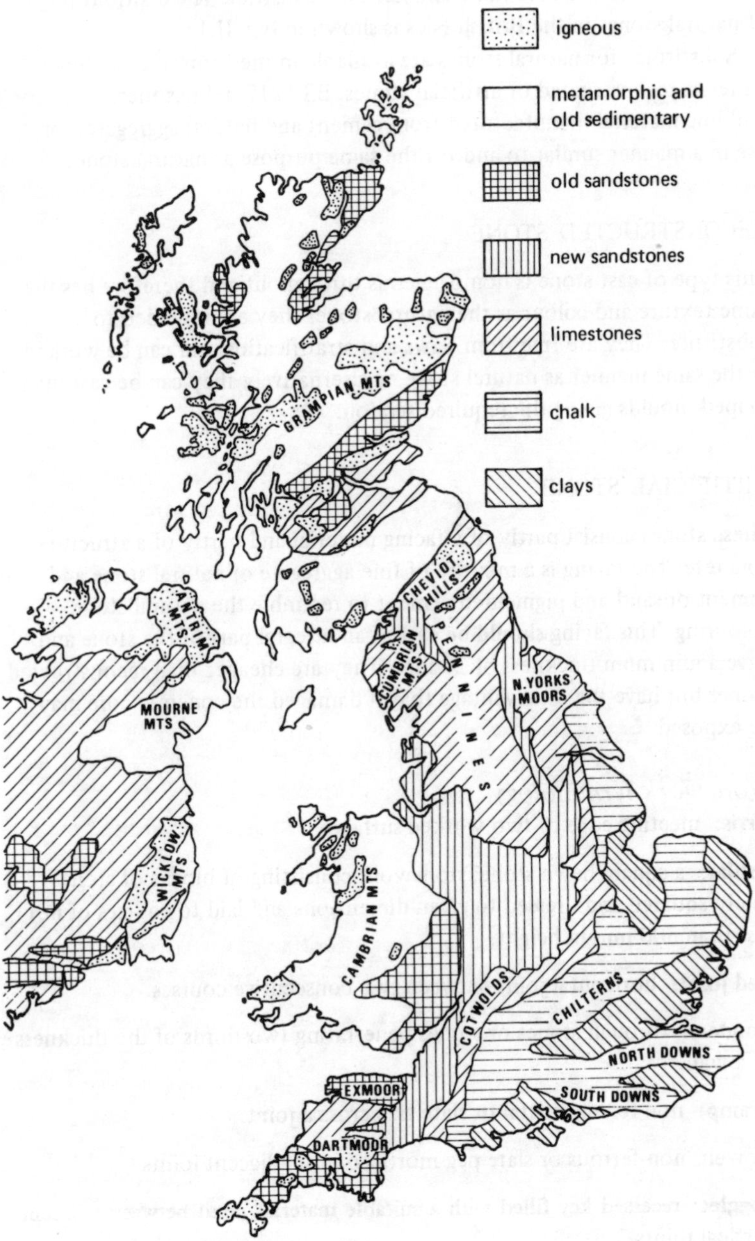
igneous
metamorphic and old sedimentary
old sandstones
new sandstones
limestones
chalk
clays
GRAMPIAN MTS
ANTRIM MTS
MOURNE MTS
CHEVIOT HILLS
CUMBRIAN MTS
PENNINES
N. YORKS MOORS
WICKLOW MTS
CAMBRIAN MTS
COTWOLDS
CHILTERNS
NORTH DOWNS
SOUTH DOWNS
EXMOOR
DARTMOOR

Fig. II.1 Distribution of natural stones

stones occur, to preserve the character of the district. The distribution of natural stones in the British Isles is shown in Fig. II.1.

Substitutes for natural stones are available in the form of cast stones either as reconstructed or artificial stones. BS 1217 defines these as a building material manufactured from cement and natural aggregate, for use in a manner similar to and for the same purpose as natural stone.

RECONSTRUCTED STONE

This type of cast stone is homogeneous throughout and therefore has the same texture and colour as the natural stones they are intended to substitute. They are free from flaws and stratification and can be worked in the same manner as natural stone or alternatively they can be cast into shaped moulds giving the required section.

ARTIFICIAL STONE

These stones consist partly of a facing material and partly of a structural concrete. The facing is a mixture of fine aggregate of natural stone and cement or sand and pigmented cement to resemble the natural stone colouring. This facing should be cast as an integral part of the stone and have a minimum thickness of 20 mm. They are cheaper than reconstructed stones but have the disadvantage that if damaged the concrete core may be exposed.

Stonework terminology

Arris: meeting edge of two worked surfaces.

Ashlar: a square hewn stone; stonework consisting of blocks of stone finely squared and dressed to given dimensions and laid to courses of not less than 300 mm in height.

Bed joint: horizontal joint between two consecutive courses.

Bonders: through stones or stones penetrating two thirds of the thickness of a wall.

Cramp: non-ferrous metal or slate tie across a joint.

Dowel: non-ferrous or slate peg morticed into adjacent joints.

Joggle: recessed key filled with a suitable material, used between adjacent vertical joints.

Lacing: course of different material to add strength.

Natural bed: plane of stratification in sedimentary stones.

Quarry sap: moisture contained in newly quarried stones.

Quoin: corner stone.

Stool: flat seating on a weathered sill for jamb or mullion.

String course: distinctive course or band used mainly for decoration.

Weathering: sloping surface to part of the structure to help shed the rain.

ASHLAR WALLING

This form of stone walling is composed of carefully worked stones, regularly coursed, bonded and set with thin or rusticated joints and is used for the majority of high-class facing work in stone. The quions are sometimes given a surface treatment to emphasise the opening or corner of the building. The majority of ashlar work is carried out in limestone varying in thickness from 100-300 mm and set in mason's putty which is a mixture of stone dust, lime putty and portland cement a typical mix ratio being 7 : 5 : 2.

Rules for ashlar work
  1. Back faces of ashlar stones should be painted with a bituminous or similar waterproofing paint.
  2. External stonework must not be taken through the thickness of the wall since this could create a passage for moisture.
  3. Ledges of cornices and external projections should be covered with lead, copper or asphalt to prevent damage by rain or birds.
  4. Moulded cornices should be raked back at 45° to counteract the cantilever action.
  5. Face of stones should be given a protective coat of slurry during construction, the slurry being washed off immediately prior to completion.

Typical details of ashlar work are shown in Figs. II.2-II.5.

RUBBLE WALLING

These are walls consisting of stones which are left in a rough or uneven state thus presenting a natural appearance to the face of the wall. These stones are usually laid with a wide joint and are frequently used in various forms in many rural areas. They can be laid dry or bedded in earth in boundary walls or bedded in lime mortar when used for the walls of farm outbuildings, if used in conjunction with ashlar stonework a cement or gauged mortar is used. It is usual to build the quoins to corners, window

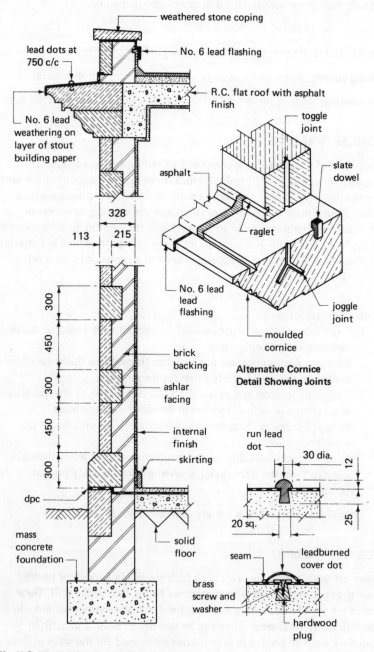

Fig. 11.2 Typical details of ashlar stonework

weathered stone coping

lead dots at 750 c/c

No. 6 lead flashing

R.C. flat roof with asphalt finish

No. 6 lead weathering on layer of stout building paper

asphalt

No. 6 lead lead flashing

toggle joint

slate dowel

raglet

joggle joint

moulded cornice

Alternative Cornice Detail Showing Joints

328

113   215

300

450

300

450

300

brick backing

ashlar facing

internal finish

skirting

dpc

mass concrete foundation

solid floor

run lead dot

30 dia.

12

20 sq.

25

seam

leadburned cover dot

brass screw and washer

hardwood plug

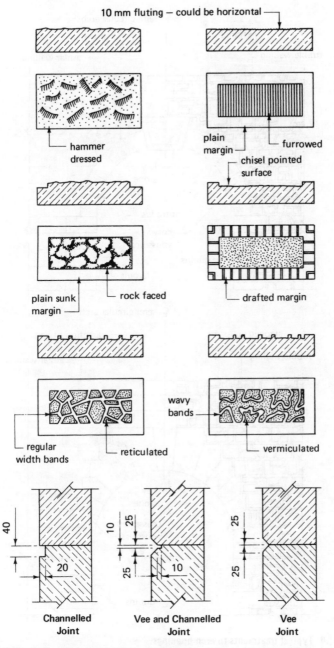

10 mm fluting — could be horizontal

hammer dressed

plain margin — furrowed

chisel pointed surface

plain sunk margin — rock faced

drafted margin

regular width bands — reticulated

wavy bands — vermiculated

Channelled Joint

Vee and Channelled Joint

Vee Joint

Fig. II.3 Typical surface and joint treatments

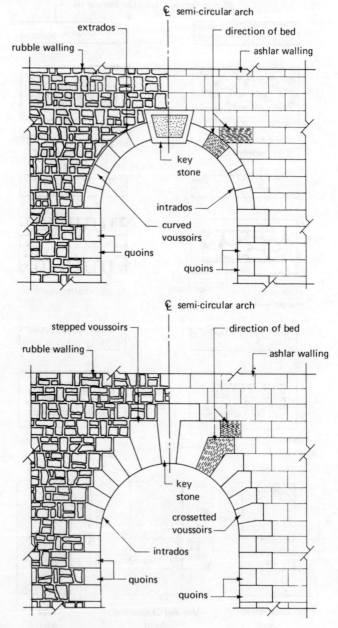

Fig. II.4 Typical treatments to arch openings

40

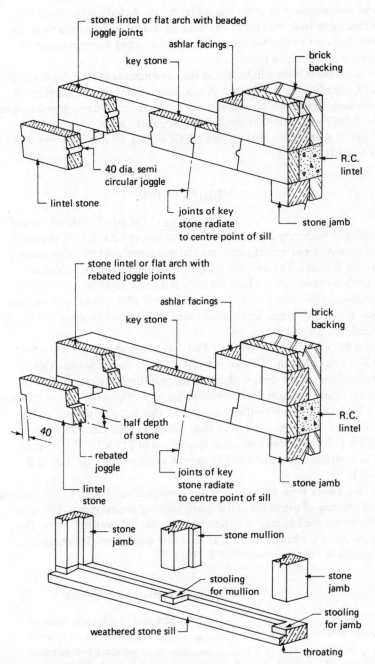

Fig. II.5 Typical treatments to square openings

Labels in figure:

stone lintel or flat arch with beaded joggle joints

ashlar facings

key stone

brick backing

40 dia. semi circular joggle

lintel stone

R.C. lintel

joints of key stone radiate to centre point of sill

stone jamb

stone lintel or flat arch with rebated joggle joints

ashlar facings

key stone

brick backing

40

half depth of stone

rebated joggle

lintel stone

R.C. lintel

joints of key stone radiate to centre point of sill

stone jamb

stone jamb

stone mullion

stooling for mullion

stone jamb

weathered stone sill

stooling for jamb

throating

and door openings in dressed or ashlar stones. As with ashlar work it is advisable to treat the face of any backing material with a suitable water-proofing coat to prevent the passage of moisture or the appearance of cement stains on the stone face.

Solid stone walls will behave in the same manner as solid brick walls with regard to the penetration of moisture and rain, it will therefore be necessary to take the same precautions in the form of damp-proof courses to comply with Part C of the Building Regulations.

Typical examples of stone and rubble walling are shown in Figs. II.4, II.6 and II.7.

BRICKWORK

The history of the art of brickmaking and the craft of bricklaying can be traced back to before 6000 B.C. It started in western Asia and spread eastwards and was introduced into this country by the Romans. The use of brickwork flourished during the third and fourth centuries, after which the craft suffered a rapid decline until its reintroduction from Flanders towards the end of the fourteenth century. Since then it has been firmly established and remains as one of the major building materials.

A brick is defined in BS 3921 Part 2 as a walling unit not exceeding 337·5 mm in length, 225 mm in width or 112·5 mm in height. This particular standard deals with bricks made of fired brickearth, clay or shale; other standards deal with those made of calcium silicate or concrete.

Bricks are known by their format size, that is the actual size plus a 10 mm joint allowance to three faces. Therefore the standard brick of 225 x 112·5 x 75 mm has actual dimensions of 215 x 102·5 x 65 mm. The terms used for bricks and brickwork are shown in Figs. II.8, II.9 and II.10.

Brickwork is used primarily in the construction of walls by the bedding and jointing of bricks into established bonding arrangements. The term also covers the building in of hollow and other lightweight blocks. The majority of the bricks used today are those made from clay or shale conforming to the requirements of BS 3921.

Manufacture of clay bricks

The basic raw material is clay, shale or brickearth of which this country has a good supply. The raw material is dug and then prepared either by weathering or grinding before being mixed with water to the right plastic condition. It is then formed into the required brick shape before being dried and fired in a kiln.

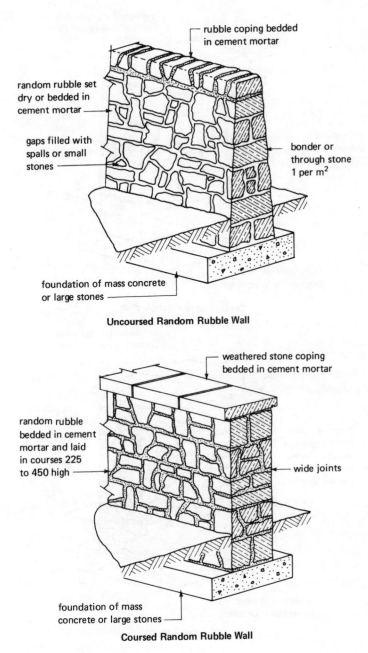

rubble coping bedded in cement mortar

random rubble set dry or bedded in cement mortar

gaps filled with spalls or small stones

bonder or through stone 1 per m²

foundation of mass concrete or large stones

Uncoursed Random Rubble Wall

weathered stone coping bedded in cement mortar

random rubble bedded in cement mortar and laid in courses 225 to 450 high

wide joints

foundation of mass concrete or large stones

Coursed Random Rubble Wall

Fig. II.6 Typical rubble walls

43

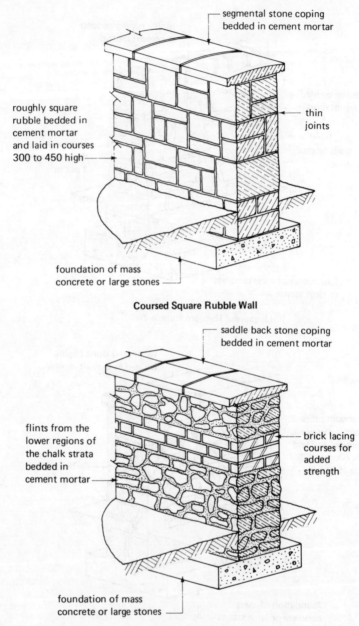

segmental stone coping
bedded in cement mortar

roughly square
rubble bedded in
cement mortar
and laid in courses
300 to 450 high

thin
joints

foundation of mass
concrete or large stones

Coursed Square Rubble Wall

saddle back stone coping
bedded in cement mortar

flints from the
lower regions of
the chalk strata
bedded in
cement mortar

brick lacing
courses for
added
strength

foundation of mass
concrete or large stones

Flint Stone Wall

Fig. II.7 Typical stone walls

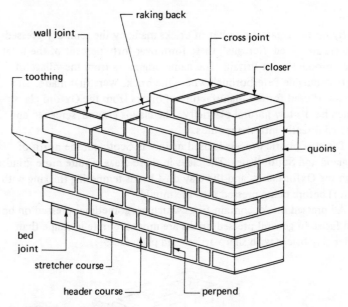

Fig. II.8 Brickwork terminology

Different clays have different characteristics, such as moisture content and chemical composition, therefore distinct variations of the broad manufacturing processes have been developed and these are easily recognised by the finished product.

PRESSED BRICKS

This type of brick is the most common used accounting for nearly two-thirds of the 8 400 million produced in this country each year. There are two processes of pressed brick manufacture they are the semi-dry and stiff plastic methods.

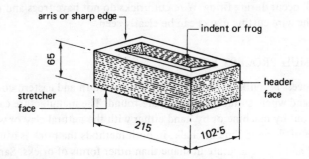

Fig. II.9 Standard brick

By far the greatest number of bricks made by the semi-dry pressed process are called 'flettons', these form over forty percent of the total brick production in Britain. The name originates from the village of Fletton outside Peterborough where the bricks were first made. This process is used for the manufacture of bricks from the Oxford clays which have a low natural plasticity. The clay is ground, screened and pressed directly into the moulds.

The stiff plastic process is used mainly in Scotland, the north of England and South Wales. The clays in these areas require more grinding than the Oxford clays and the clay dust needs tempering (mixing with water) before being pressed into the mould.

All pressed bricks contain frogs which are sometimes pressed on both bed faces. In general pressed bricks are more accurate in shape than other clay bricks with sharp arrises and plain faces.

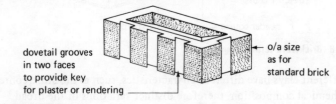

dovetail grooves
in two faces
to provide key
for plaster or rendering

o/a size
as for
standard brick

Fig. II.10 Keyed brick

WIRE CUT BRICKS

Approximately 28% of bricks produced in Britain are made by this process. The clay which is usually fairly soft and of a fine texture is extruded as a continuous ribbon and is cut into brick units by tightly stretched wires spaced at the height or depth for the required brick. Allowance is made during the extrusion and cutting for the shrinkage that will occur during firing. Wire cut bricks do not have frogs and on many the wire cutting marks can be clearly seen.

SOFT MUD PROCESS BRICKS

This process is confined mainly to the south-eastern and eastern counties of England where suitable soft clays are found. The manufacture can be carried out by machine or by hand either with the natural clay or with a mixture of clay and lime or chalk. In both methods the brick is usually frogged and is less accurate in shape than other forms of bricks. Sand is usually used in the moulds, to enable the bricks to be easily removed, and this causes an uneven patterning or creasing on the face.

Brick classification

No standard system for the classification of bricks has yet been devised, bricks are generally known by the terms given in BS 3921 or by the description given by the brick manufacturer or a combination of the two.

BS 3921, PART 2

This standard gives three headings:

1. Varieties

Common: suitable for general building work but having no special claim to give an attractive appearance.

Facing: specially made or selected to have an attractive appearance when used without rendering or plaster.

Engineering: having a dense and strong semi-vitreous body conforming to defined limits for absorption and strength.

2. Qualities

Internal: suitable for internal use only, may need protection on site during bad weather or during the winter.

Ordinary: less durable than special quality but normally durable in the external face of a building. Some types are unsuitable for exposed situations.

Special: for use in conditions of extreme exposure where the structure may become saturated and frozen such as retaining walls and pavings.

3. Types

Solid: those in which small holes passing through or nearly through the brick do not exceed 25% of its volume or in which frogs do not exceed 20% of its volume. A small hole is defined as a hole less than 20 mm wide or less than 500 mm^2 in area.

Perforated: those in which holes passing through the brick exceed 25% of its volume and the holes are small as defined above.

Hollow: those in which the holes passing through the brick exceed 25% of its volume and the holes are larger than those defined as small holes.

Cellular: those in which the holes are closed at one end and exceed 20% of the volume of the brick.

Bricks may also be classified by one or more of the following:
Place of origin, for example, London.
Raw material, for example, clay.
Manufacture, for example, wire cut.
Use, for example, foundation.
Colour, for example, blue.
Surface texture, for example, sand-faced.

Calcium silicate bricks

These bricks are also called sandlime and sometimes flintlime bricks and are covered by BS 187, Part 2, which gives eight classes of brick, the higher the numbered class the stronger is the brick. The format size of a calcium silicate brick is the same as that given for a standard clay brick.

These bricks are made from carefully selected clean sand and/or crushed flint mixed with controlled quantities of lime and water. At this stage colouring pigments can be added if required, the relatively dry mix is then fed into presses to be formed into the required shape. The moulded bricks are then hardened in sealed and steam pressurised autoclaves. This process, which takes from seven to ten hours, causes a reaction between the sand and the lime resulting in a strong homogenous brick which is ready for immediate delivery and laying. The bricks are very accurate in size and shape but do not have the individual character of clay bricks.

Concrete bricks

These are made from a mixture of inert aggregate and cement in a similar fashion to calcium silicate bricks and are cured either by natural weathering or in an autoclave. Details of the types and properties available as standard concrete bricks are given in BS 6073.

Mortars for brickwork

The mortar used in brickwork transfers the stresses, tensile, compressive and shear uniformly between adjacent bricks. To do this it must satisfy certain requirements:

1. Have adequate strength, but not greater than that required for the design strength.
2. Have good workability.
3. Needs to retain plasticity long enough for the bricks to be laid.

4. Must be durable over a long period.
5. Bond well to the bricks.
6. Should be able to be produced at an economic cost.

If the mortar is weaker than the bricks shrinkage cracks will tend to follow the joints of the brickwork and these are reasonably easy to make good. If the mortar is stronger than the bricks shrinkage cracks will tend to be vertical through the joints and the bricks thus weakening the fabric of the structure.

MORTAR MIXES

Mortar is a mixture of sand and lime or a mixture of sand and cement with or without lime. Proportioning of the materials can be carried by volume but this method is inaccurate and it is much better to proportion by weight. The effect of the lime is to make the mix more workable, but as the lime content increases the mortar's resistance to damage by frost action decreases.

Plasticisers by having the effect of entraining small bubbles of air in the mix and breaking down surface tension will also increase the workability of a mortar.

Mortars should never be re-tempered and should be used within two hours of mixing or be discarded.

Typical mixes (by volume)

Cement mortar 1 : 3 (cement : sand) suitable for brickwork in exposed conditions such as parapets and for brickwork in foundations.

Lime mortar 1 : 3 (lime : sand) for internal use only.

Gauged mortars (cement : lime : sand):
1 : 1 : 6 suitable for most conditions of severe exposure.
1 : 2 : 9 suitable for most conditions except those of severe exposure.
1 : 3 : 12 internal use only.

Dampness penetration

It is possible for dampness to penetrate into a building through the brick walls by one or more of three ways:

1. By the rain penetrating the head of the wall and soaking down into the building below the roof level.
2. By the rain beating against the external wall and soaking through the fabric into the building.
3. By ground moisture entering the wall at or near to the base and

creeping up the wall by capillary action and entering the building above the ground floor level.

Nos. 1 and 3 can be overcome by the insertion of a suitable damp-proof course in the thickness of the wall. 2 can be overcome by one of two methods:

(a) Applying to the exposed face of the wall a barrier such as cement rendering or some suitable cladding like vertical tile hanging.

(b) by constructing a cavity wall, whereby only the external skin becomes damp, the cavity providing a suitable barrier to the passage of moisture through the wall.

DAMP-PROOF COURSES

The purpose of a damp-proof course in a building is to provide a barrier to the passage of moisture from an external source into the fabric of the building or from one part of the structure to another. Damp-proof courses may be either horizontal or vertical and can generally be divided into three types:

1. Those below ground level to prevent the entry of moisture from the soil.
2. Those placed just above ground level to prevent moisture creeping up the wall by capillary action, this is sometimes called rising damp.
3. Those placed at openings, parapets and similar locations to exclude the entry of the rainwater which falls directly onto fabric of the structure.

Materials for damp-proof courses

BS 743 gives seven suitable materials for the construction of damp-proof courses all of which should have the following properties:

(a) Completely impervious.

(b) Durable, having a longer life than the other components in the building and therefore should not need replacing during its lifetime.

(c) Be in comparatively thin sheets so as to prevent disfigurement of the building.

(d) Be strong enough to support the loads placed upon it without exuding from the wall.

(e) Be flexible enough to give with any settlement of the building without fracturing.

Lead: this should be at least code No. 4—BS 1178 lead, it is a flexible material supplied in thin sheets and therefore large irregular shapes with few joints can be formed but it has the disadvantages of being expensive and liable to exude under heavy loadings.

Copper: should have a minimum thickness of 0·25 mm, like lead it is supplied in thin sheets and is expensive.

Bitumen: this is supplied in the form of a felt usually to brick widths and is therefore laid quickly with the minimum number of joints. Various bases are available, such as hessian, fibre, asbestos and lead, all of which are inexpensive but have the main disadvantage of being easily torn.

Mastic asphalt: applied in two layers giving a total thickness of 25 mm, it is applied *in situ* and is therefore jointless but is expensive in small quantities.

Polythene: black low density polythene sheet of single thickness not less than 0·5 mm thick should be used, it is easily laid but can be torn and punctured easily.

Slates: these should not be less than 230 mm long nor less than 4 mm thick and laid in two courses set breaking the joint in cement mortar 1 : 3. Slates have limited flexibility but are impervious and very durable, cost depends upon the area in which the building is being erected.

Bricks: should comply with the requirements of BS 3921 and be laid in two courses in cement mortar 1 : 3, can be out of context with the general façade of the building.

Brickwork bonding

When building with bricks it is necessary to lay the bricks to some recognised pattern or bond in order to ensure stability of the structure and to produce a pleasing appearance. All the various bonds are designed so that no vertical joint in any one course is directly above or below a vertical joint in the adjoining course. To simplify this requirement special bricks are produced or cut from whole bricks on site, a selection of these special bricks is shown in Fig. II.11. The various bonds are also planned to give the greatest practical amount of lap to all the bricks and this should not be less than a quarter of a brick length. Properly bonded brickwork distributes the load over as large an area of brickwork as possible, then angle of spread of the load through the bonded brickwork is 60°.

Common bonds
Stretcher bond: consists of all stretchers in every course and is used for

half brick walls and the half brick skins of hollow or cavity walls (see Fig. II.12).

English bond: a very strong bond consisting of alternate courses of headers and stretchers (see Fig. II.13).

Flemish bond: each course consists of alternate headers and stretchers, its appearance is considered to be better than English bond but it is not quite so strong. This bond requires fewer facing bricks than English bond needing only 79 bricks per square metre as opposed to 89 facing bricks per square metre for English bond. This bond is sometimes referred to as double Flemish bond (see Fig. II.14).

Single Flemish bond: a combination of English and Flemish bonds having Flemish bond on the front face with a backing of English bond. It is considered to be slightly stronger than Flemish bond. The thinnest wall that can be built using this bond is a one and half brick wall.

English garden wall bond: consists of three courses of stretchers to one course of headers.

Flemish garden wall bond: consists of one header to every three stretchers in every course, this bond is fairly economical in facing bricks and has a pleasing appearance.

Special bonds

Rat-trap bond: this is a brick on edge bond and gives a saving on materials and loadings, suitable as a backing wall to a cladding such as tile hanging (see Fig. II.15).

Quetta bond: used on one and half brick walls for added strength, suitable for retaining walls (see Fig. II.15).

METRIC MODULAR BRICKWORK

The standard format brick does not fit reasonably well into the system of dimensional coordination with its preferred dimension of 300 mm, therefore metric modular bricks have been designed and produced with four different formats (see Fig. II.16).

The bond arrangements are similar to the well-known bonds but are based on third bonding, that is the overlap is one-third of a brick and not one-quarter as with the standard format brick. Examples of metric modular brick bonding are shown in Fig. II.16.

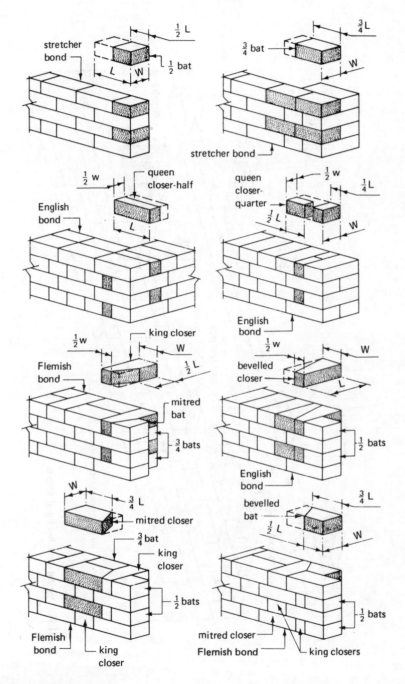

Fig. II.11 Special bricks

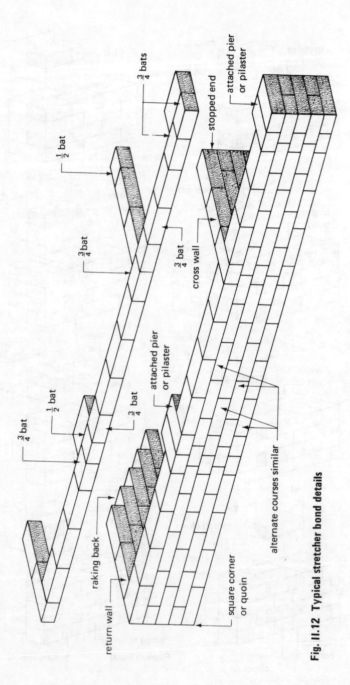

Fig. II.12 Typical stretcher bond details

$\frac{3}{4}$ bats

$\frac{1}{2}$ bat

$\frac{3}{4}$ bat

$\frac{1}{2}$ bat

$\frac{3}{4}$ bat

$\frac{1}{2}$ bat

$\frac{3}{4}$ bat

stopped end

attached pier or pilaster

cross wall

attached pier or pilaster

raking back

return wall

square corner or quoin

alternate courses similar

54

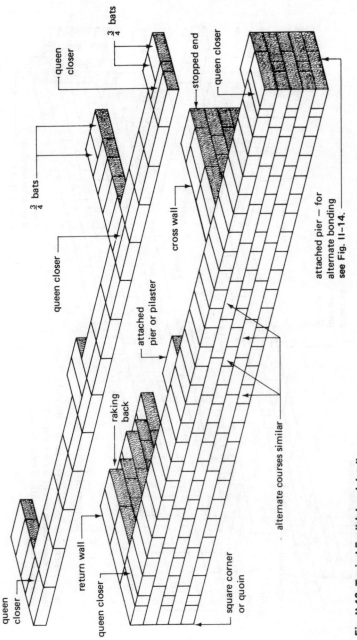

Fig. II.13 Typical English bond details

queen closer

return wall

queen closer

raking back

attached pier or pilaster

cross wall

square corner or quoin

alternate courses similar

queen closer

¾ bats

queen closer

¾ bats

stopped end

queen closer

attached pier — for alternate bonding see Fig. II–14.

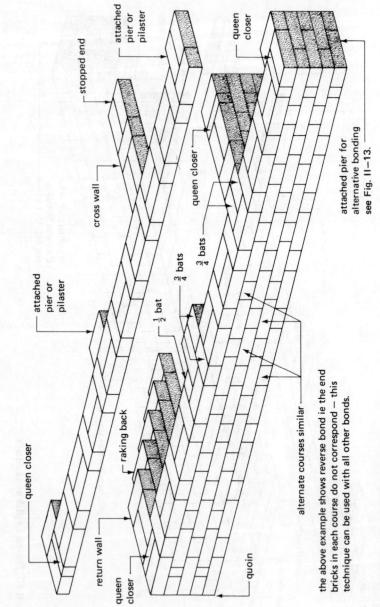

the above example shows reverse bond ie the end bricks in each course do not correspond — this technique can be used with all other bonds.

Fig. II.14 Typical Flemish bond details

queen closer

attached pier or pilaster

stopped end

cross wall

attached pier or pilaster

return wall

queen closer

raking back

$\frac{1}{2}$ bat

$\frac{3}{4}$ bats

$\frac{3}{4}$ bats

queen closer

quoin

alternate courses similar

queen closer

attached pier for alternative bonding see Fig. II–13.

56

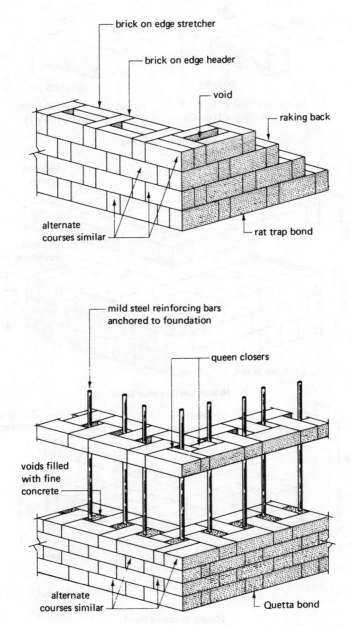

brick on edge stretcher

brick on edge header

void

raking back

alternate courses similar

rat trap bond

mild steel reinforcing bars anchored to foundation

queen closers

voids filled with fine concrete

alternate courses similar

Quetta bond

Fig. II.15 Special bonds

57

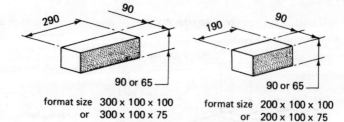

format size   300 x 100 x 100
      or   300 x 100 x 75

format size   200 x 100 x 100
      or   200 x 100 x 75

Metric Modular Bricks

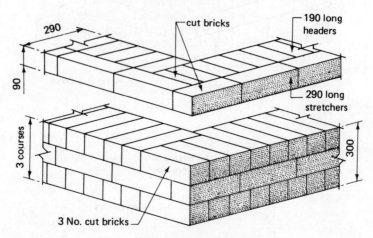

Header and Stretcher Bond

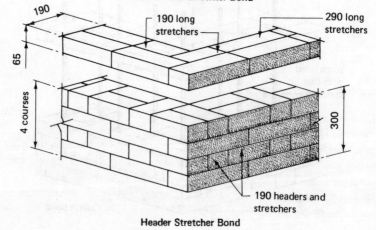

Header Stretcher Bond

Fig. II.16 Metric modular brickwork

FOOTINGS

These are wide courses of bricks placed at the base of a wall to spread the load over a greater area of the foundations. This method is seldom used today, instead the concrete foundation would be reinforced to act as a beam or reinforced strip. The courses in footings are always laid as headers as far as possible, stretchers if needed are laid in the centre of the wall (see Fig. II.17).

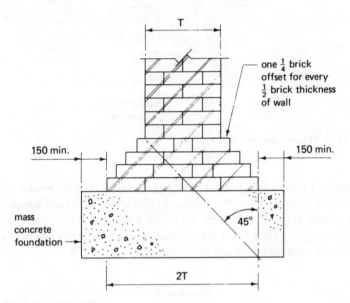

one $\frac{1}{4}$ brick offset for every $\frac{1}{2}$ brick thickness of wall

T

150 min.   150 min.

mass concrete foundation

45°

2T

Fig. II.17 Typical footings

BOUNDARY WALLS

These are subjected to severe weather conditions and therefore should be correctly designed and constructed. If these walls are also acting as a retaining wall the conditions will be even more extreme, but the main design principle of the exclusion of water remains the same. The presence of water in brickwork can lead to frost damage, mortar failure and efflorescence. The incorporation of adequate damp-proof courses and overhanging throated copings is of the utmost importance in this form of structure (see Fig. II.18).

Efflorescence
This is a white stain appearing on the face of brickwork caused by

59

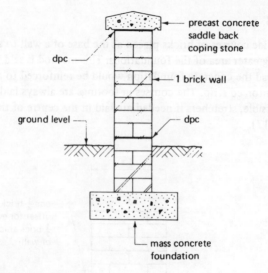

precast concrete
saddle back
coping stone

dpc

1 brick wall

ground level

dpc

mass concrete
foundation

Fig. II.18 Typical boundary wall

deposits of soluble salts formed on or near the surface of the brickwork as a result of evaporation of the water in which they have been dissolved. It is usually harmless and disappears within a short period of time; dry brushing or with clean water may be used to remove the salt deposit but the use of acids should be left to the expert.

BLOCKWORK

A block can be defined as a walling unit exceeding the dimensions specified for bricks given in BS 3921 and that its height shall not exceed either its length or six times its thickness to avoid confusion with slabs or panels. Blocks are produced from clay, precast concrete and aerated concrete.

Clay Blocks

These are covered by BS 3921 which gives a format size of 300 x 225 x 62·5; 75, 100 or 150 mm wide. These blocks, which are hollow, are made by an extrusion process and fired as for clay bricks. The standard six cavity block is used mainly for the inner skin of a cavity wall, whereas the three cavity block is primarily intended for partition work. Special corner, closer, fixing and conduit blocks are produced to give the range good flexibility in design and lay-out. Typical details are shown in Fig. II.19.

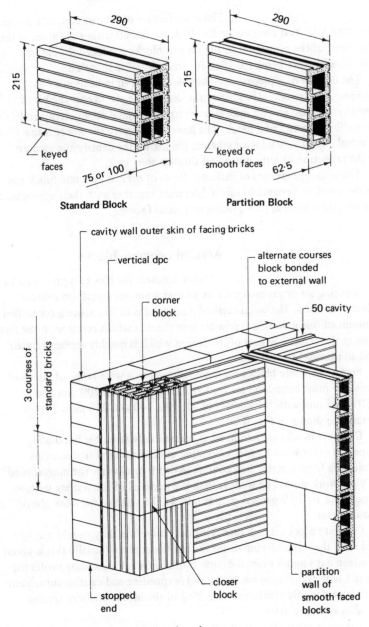

Fig. II.19 Hollow clay blocks and blockwork

61

Precast concrete blocks

The manufacture of Precast concrete blocks or masonry units is covered by BS 6073. No classifications are given in this standard but the properties of the various blocks produced should be considered before being specified for any particular situation.

The density of a precast concrete block gives an indication of its compressive strength—the greater the density the stronger is the block. Density will also give an indication as to the thermal conductivity and acoustic properties of a block. The lower the density the lower is the thermal conductivity factor whereas the higher the density the greater is the reduction of airborne sound through the block.

The actual properties of different types of precast concrete block can be obtained from manufacturers' literature together with their appearance classification such as plain, facing or special facing.

Aerated concrete blocks

Aerated concrete for blocks is produced by introducing air or gas into the mix so that when set a uniform cellular block is formed. The usual method employed is to introduce a controlled amount of fine aluminium powder into the mix which reacts with the free lime in the cement to give off hydrogen which is quickly replaced by air and so provides the aeration.

Precast concrete blocks are manufactured to a wide range of standard sizes, the most common face format sizes being 400 x 200 mm and 450 x 225 mm with thicknesses of 75, 100, 140 and 215 mm. Typical details are shown in Fig. II.20.

Concrete blocks are laid in what is essentially stretcher bond and joined to other walls by block bonding or leaving metal ties or strips projecting from suitable bed courses. As with brickwork the mortar used in blockwork should be weaker than the material of the walling unit, generally a 1 : 2 : 9 gauged mortar mix will be suitable for work above ground level.

Concrete blocks shrink on drying out, therefore they should not be laid until the initial drying shrinkage has taken place (usually this is about fourteen days under normal drying conditions) and should be protected on site to prevent them becoming wet, expanding and causing subsequent shrinkage possibly resulting in cracking of the blocks and any applied finishes such as plaster.

The main advantages of blockwork over brickwork are:

1. Labour saving—easy to cut, larger units.

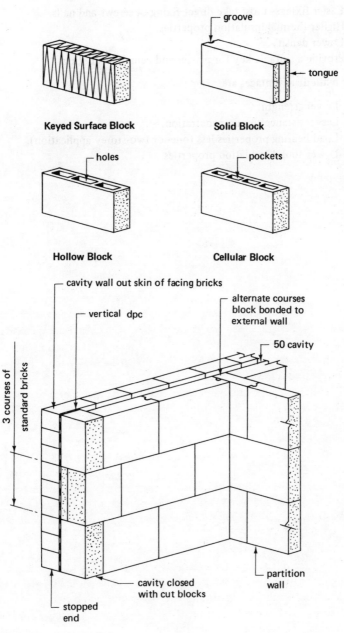

Keyed Surface Block

Solid Block

groove

tongue

Hollow Block

holes

Cellular Block

pockets

cavity wall out skin of facing bricks

vertical dpc

alternate courses block bonded to external wall

50 cavity

3 courses of standard bricks

cavity closed with cut blocks

partition wall

stopped end

Fig. 11.20 Precast concrete blocks and blockwork

2. Easier fixings—most take direct fixing of screws and nails.
3. Higher thermal insulation properties.
4. Lower density.
5. Provide a suitable key for plaster and cement rendering.

The main disadvantages are:

(*a*) Lower strength.
(*b*) Less resistance to rain penetration.
(*c*) Load bearing properties less (one- or two-storey application).
(*d*) Lower sound insulation properties.

5
Cavity walls

A wall constructed in two leaves or skins with a space or cavity between them is called a cavity wall and it is the most common form of external wall used in domestic building today. The main purpose of constructing a cavity wall is to prevent the penetration of rain to the internal surface of the wall. It is essential that the cavity is not bridged in any way as this would provide a passage for the moisture.

Air bricks are sometimes used to ventilate the cavity and these would be built in at the head and base of the cavity wall in order that a flow of air may pass through the cavity thus drying out any moisture that has penetrated the outer leaf. Unless the wall is exposed to very wet conditions the practice of inserting air bricks to ventilate the cavity is not recommended since it lowers the thermal and sound insulation values of the wall.

The main consideration in the construction of a cavity wall above ground level damp-proof course is the choice of a brick or block which will give the required durability, strength and appearance and also conform to Building Regulation requirements. The main function of the wall below ground level damp-proof course is to transmit the load safely to the foundations, in this context the two half-brick leaves forming the wall act as retaining walls. There is a tendency for the two leaves to move towards each other, due to the pressure of the soil, and the space provided by the cavity. To overcome this problem it is common practice to fill the cavity below ground level with a weak mix of concrete thus creating a 'solid' wall in the ground (see Fig. II.22). It is also advisable to leave out every fourth vertical joint in the external leaf at the base of the cavity and

65

above the cavity fill, to allow any moisture trapped in the cavity a means of escape.

Parapets, whether solid or cavity construction, are exposed to the elements on three sides and need careful design and construction. They must be provided with adequate barriers to moisture in the form of damp-proof courses since dampness could penetrate the structure by soaking down the wall and by-passing the roof and entering the building below the uppermost ceiling level. A solid parapet wall should not be less than 150 mm thick and not less than the thickness of the wall on which it is carried and its height should not exceed four times its thickness. The recommended maximum heights of cavity wall parapets is shown in Fig. II.23.

BUILDING REGULATIONS 1985

Regulation A1 requires that a building shall be so constructed that the combined dead, imposed and wind loads are sustained and transmitted to the ground safely and without causing any movement which will impair the stability of any part of another building. Guidance to meet the above requirements for cavity walls is given in Approved Document A.

Part C of this document deals with full storey height cavity walls for residential buildings of up to three storeys and recommends that:

1. The unit compressive strengths of bricks and blocks should be 5 N/mm² and 2.8 N/mm² respectively.
2. Cavities should be not less than 50 mm nor more than 100 mm in width at any level.
3. Wall ties should comply with BS 1243 or other not less suitable type and should have the maximum spacings given in Table C3. Fig. II.21 gives the maximum wall tie spacings for a cavity width of 50 to 75 mm.
4. All cavity walls should have leaves at least 90 mm thick.
5. The combined thickness of the two leaves of a cavity wall plus 10 mm should not be less than the thickness required for a solid wall of the same length and height (for a maximum wall length of 9.000 with a height between 3.500 and 9.000 the two leaves plus 10 mm is 200 mm for the whole of its height—see Table C2 in AD. A).
6. Mortar should be as given for mortar designation (iii) in BS 5628 Part 1 or a gauged mortar mix of 1 : 1 : 6 by volume.
7. Cavity walls of any length need to be provided with roof lateral support and those over 3.000 in length will also require floor lateral support at every floor forming a junction with the supported wall.

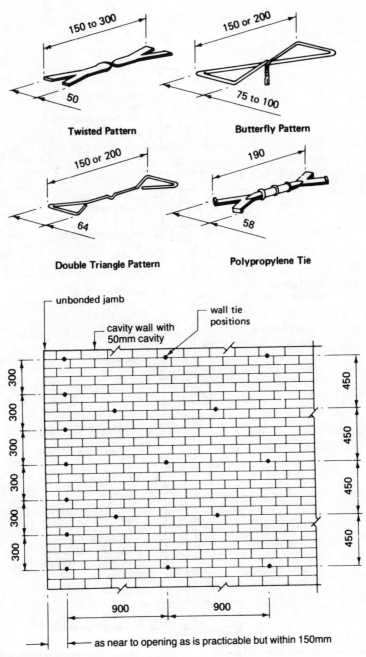

150 to 300

50

Twisted Pattern

150 or 200

75 to 100

Butterfly Pattern

150 or 200

64

Double Triangle Pattern

190

58

Polypropylene Tie

unbonded jamb

cavity wall with
50mm cavity

wall tie
positions

300

300

300

300

300

300

450

450

450

450

900

900

as near to opening as is practicable but within 150mm

Fig. II.21 Wall ties

67

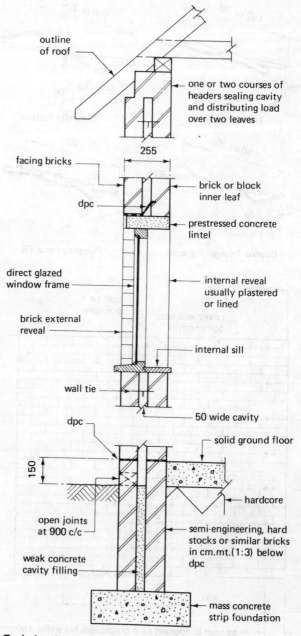

outline of roof

one or two courses of headers sealing cavity and distributing load over two leaves

255

facing bricks

dpc

brick or block inner leaf

prestressed concrete lintel

direct glazed window frame

internal reveal usually plastered or lined

brick external reveal

internal sill

wall tie

50 wide cavity

dpc

solid ground floor

150

hardcore

open joints at 900 c/c

semi-engineering, hard stocks or similar bricks in cm.mt.(1:3) below dpc

weak concrete cavity filling

mass concrete strip foundation

Fig. II.22 Typical cavity wall details

If roof lateral support is not provided by type of covering (tiles or slates), a pitch of 15° or more plus a minimum bearing of 75 mm then durable metal straps with a minimum cross-section of 30 mm × 5 mm will be needed at not more than 2.000 centres. If the floor does not have at least a 90 mm bearing on the supported wall lateral support should be provided by similar straps at not more than 2.000 centres or the joists should be fixed using restraint-type joist hangers.

PREVENTION OF DAMP IN CAVITY WALLS

Approved Document C recommends a cavity to be carried down at least 150 mm below the lowest damp-proof course and that any bridging of the cavity, other than a wall tie or closing course protected by the roof, is to have a suitable damp-proof course to prevent the passage of moisture across the cavity. Where the cavity is closed at the jambs of openings a vertical damp-proof course should be inserted unless some other suitable method is used to prevent the passage of moisture from the outer leaf to the inner leaf of the wall.

Approved Document C recommends a damp-proof course to be inserted in all external walls at least 150 mm above the highest adjoining ground or paving to prevent the passage of moisture rising up the wall and into the building, unless the design is such that the wall is protected or sheltered.

ADVANTAGES OF CAVITY WALL CONSTRUCTION

These can be listed as follows:

(a) Able to withstand a driving rain in all situations from penetrating to the inner wall surface.
(b) Gives good thermal insulation, keeping the building warm in winter and cool in the summer.
(c) No need for external rendering.
(d) Enables the use of cheaper and alternative materials for the inner construction.
(e) A nominal 255 mm cavity wall has a higher sound insulation value than a standard one brick thick wall.

DISADVANTAGES OF CAVITY WALL CONSTRUCTION

These can be listed as follows:

1. Requires a high standard of design and workmanship to produce a soundly constructed wall; this will require good supervision during construction.
2. The need to include a vertical damp-proof course to all openings.
3. Slightly dearer in cost than a standard one brick thick wall.

max heights:
total sum of leaf thicknesses
≤ 200 max 600
> 200 ≤ max 860

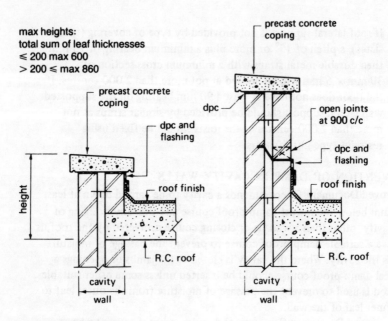

Fig. 11.23 Parapets

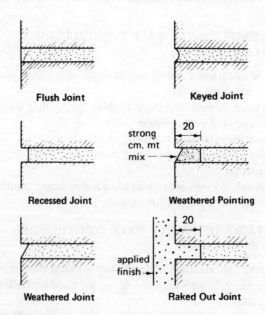

Flush Joint

Keyed Joint

Recessed Joint

strong
cm. mt
mix →

20

Weathered Pointing

Weathered Joint

applied
finish

20

Raked Out Joint

Fig. 11.24 Brick joints

Jointing and pointing

These terms are used for the finish given to both the vertical and horizontal joints in brickwork irrespective of whether the wall is of brick, block, solid or cavity construction.

Jointing is the finish given to the joints when carried out as the work proceeds.

Pointing is the finish given to the joints by raking out to depth of approximately 20 mm and filling in on the face with a hard setting cement mortar which could have a colour additive. This process can be applied to both new and old buildings. Typical examples of jointing and pointing are shown in Fig. II.24.

6
Openings in walls

An opening in an external wall consists of a head, jambs or reveals and a sill or threshold.

Head

The function of a head is to carry the triangular load of brickwork over the opening and transmit this load to the jambs at the sides. To fulfil this task it must have the capacity to support the load without unacceptable deflection. A variety of materials and methods is available in the form of a lintel or beam such as:

Timber: suitable for light loads and small spans, the timber should be treated with a preservative to prevent attack by beetles or fungi.

Steel:
For small openings—a mild steel flat or angle section can be used to carry the outside leaf of a cavity wall, the inner leaf being supported by a concrete or steel lintel.
For medium spans—a channel or joist section is usually suitable.
For large spans—a universal beam section to design calculations will be needed.
   Steel lintels which are exposed to the elements should be either galvanised or painted with several coats of bituminous paint to give them protection against corrosion.

Concrete: these can be designed as *in situ* or precast reinforced beams or lintels and can be used for all spans. Prestressed concrete lintels are available for the small and medium spans.

Stone: these can be natural, artificial or reconstructed stone but are generally used as a facing to a steel or concrete lintel (see Fig. II.5).

Brick: unless reinforced with mild steel bars or mesh, brick lintels are only suitable for small spans up to 1 m, but like stone, bricks are also employed as a facing to a steel or concrete lintel.

Lintels require a bearing at each end of the opening, the amount will vary with the span but generally it will be 100 mm for the small spans and up to 225 mm for the medium and large spans. In cavity walling a damp-proof course will be required where the cavity is bridged by the lintel and this should extend at least 150 mm beyond each end of the lintel. Open joints are sometimes used to act as weep holes; these are placed at 900 mm centres in the outer leaf immediately above the damp-proof course. Typical examples of head treatments to openings are shown in Fig. II.25.

Arches

These are arrangements of wedged shaped bricks designed to support each other and carry the load over the opening round a curved profile to abutments on either side—full details of arch construction are given in the next chapter.

Jambs

In solid walls these are bonded to give the required profile and strength; examples of bonded jambs are shown in Fig. II.11. In cavity walls the cavity can be closed at the opening by using a suitable frame or by turning one of the leaves towards the other forming a butt joint in which is incorporated a vertical damp-proof course as required by the Building Regulations. Typical examples of jamb treatments to openings are shown in Fig. II.26.

Sill

The function of a sill is to shed the rain water, which has run down the face of the window or door and collected at the base, away from the opening and the face of the wall. Many methods and materials are available; appearance and durability are the main requirements since a sill is not a member which is needed to carry heavy loads. Sills, unlike lintels, do not require a bearing at each end. Typical examples of sill treatments to openings are shown in Fig. II.27

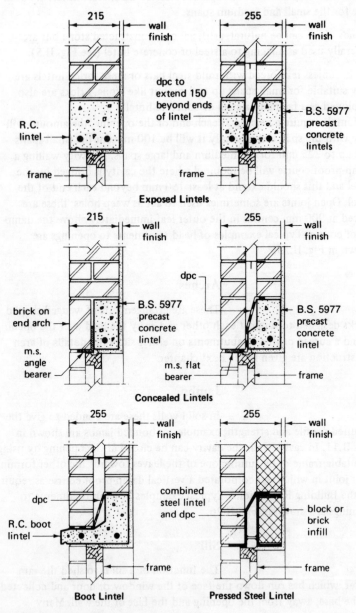

Fig. 11.25 Typical head treatments to openings

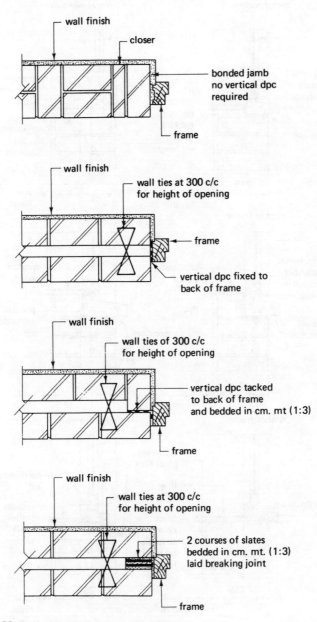

wall finish

closer

bonded jamb
no vertical dpc
required

frame

wall finish

wall ties at 300 c/c
for height of opening

frame

vertical dpc fixed to
back of frame

wall finish

wall ties of 300 c/c
for height of opening

vertical dpc tacked
to back of frame
and bedded in cm. mt (1:3)

frame

wall finish

wall ties at 300 c/c
for height of opening

2 courses of slates
bedded in cm. mt. (1:3)
laid breaking joint

frame

Fig. II.26 Typical jamb treatments to openings

75

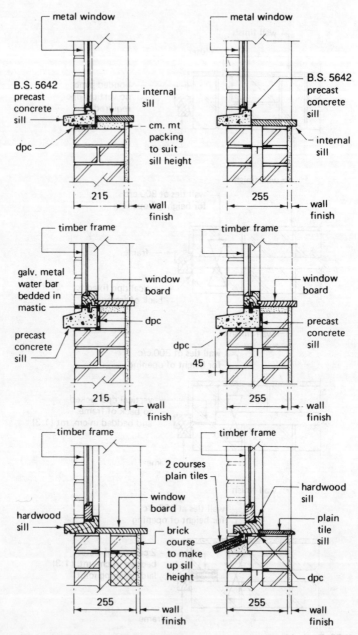

Fig. II.27 Typical sill treatments to openings

7
Arches

These are arrangements of wedge shaped
bricks called 'voussoirs' which are designed to support each other and
carry the load over the opening, round a curved profile, to abutments on
either side. An exception to this form is the flat or 'soldier' arch constructed of bricks laid on end or on edge.

When constructing an arch it must be given temporary support until
the brick joints have set and the arch has gained sufficient strength to
support itself and carry the load over the opening. These temporary
supports are called centres and are usually made of timber; their design
is governed by the span, load and thickness of the arch to be constructed.

Soldier arches

This type of arch consists of a row of bricks
showing on the face either the end or the edge of the bricks. Soldier arches
have no real strength and if the span is over 1 000 mm they will require
some form of permanent support such as a metal flat or angle (see
Fig. II.25). If permanent support is not given the load will be transferred
to the head of the frame in the opening instead of the jambs on either
side. Small spans can have an arch of bonded brickwork by inserting into
the horizontal joints immediately above the opening some form of
reinforcement such as expanded metal or bricktor which is a woven strip
of high tensile steel wires designed for the reinforcement of brick and
stone walls. It is also possible to construct a soldier arch by inserting

77

metal cramps in the vertical joints and casting these into an *in situ* backing lintel of reinforced concrete.

Rough arches

These arches are constructed of ordinary uncut bricks and being rectangular in shape they give rise to wedge shaped joints. To prevent the thick end of the joint from becoming too excessive rough arches are usually constructed in header courses. The rough arch is used mainly as a backing or relieving arch to a gauged brick or stone arch but they are sometimes used in facework for the cheaper form of building or where appearance is of little importance.

Gauged arches

These are the true arches and are constructed of bricks cut to the required wedge shape called voussoirs. The purpose of voussoirs is to produce a uniform thin joint which converges on to the centre point or points of the arch. There are two methods of cutting the bricks to the required wedge shape, namely, axed and rubbed. If the brick is of a hard nature it is first marked with a tin saw, to produce a sharp arris, and then it is axed to the required profile. For rubbed brick arches a soft brick called a rubber is used; the bricks are first cut to the approximate shape with a saw and are then finished off with an abrasive stone or file to produce the sharp arris. In both cases a template of plywood or hardboard to the required shape will be necessary for marking out the voussoirs. Typical examples of stone arches are shown in Fig. II.4; the terminology and setting out of simple brick arches is shown in Fig. II.28.

CENTRES

These are temporary structures, usually of light timber construction, which are strong enough to fulfil their function of supporting arches of brick or stone while they are being built and until they are sufficiently set to support themselves and the load over the opening. Centres can be an expensive item to a builder, therefore their design should be simple and adaptable so that as many uses as possible can be obtained from any one centre. A centre is always less in width than the soffit of an arch to allow for plumbing, that is, alignment and verticality of the face with a level or rule

78

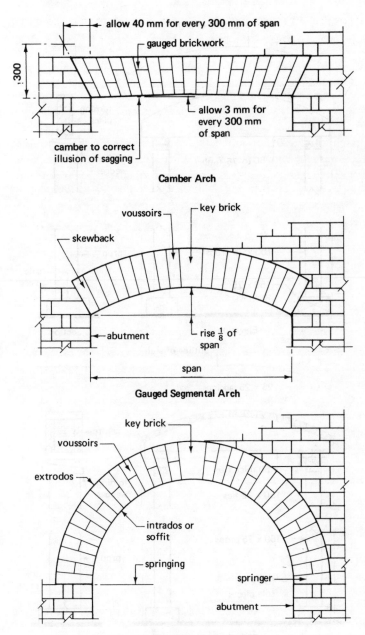

allow 40 mm for every 300 mm of span

gauged brickwork

300

allow 3 mm for every 300 mm of span

camber to correct illusion of sagging

Camber Arch

voussoirs

key brick

skewback

abutment

rise $\frac{1}{8}$ of span

span

Gauged Segmental Arch

key brick

voussoirs

extrodos

intrados or soffit

springing

springer

abutment

Gauged Semi-circular Arch

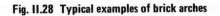

Fig. II.28 Typical examples of brick arches

79

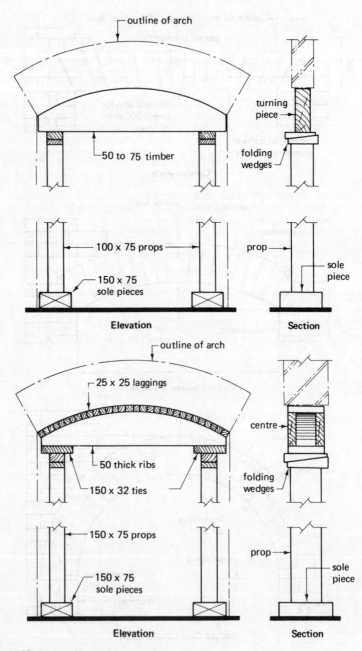

outline of arch

turning piece

folding wedges

50 to 75 timber

100 x 75 props

prop

sole piece

150 x 75 sole pieces

Elevation

Section

outline of arch

25 x 25 laggings

centre

folding wedges

50 thick ribs

150 x 32 ties

150 x 75 props

prop

sole piece

150 x 75 sole pieces

Elevation

Section

Fig. II.29 Centres for small span arches

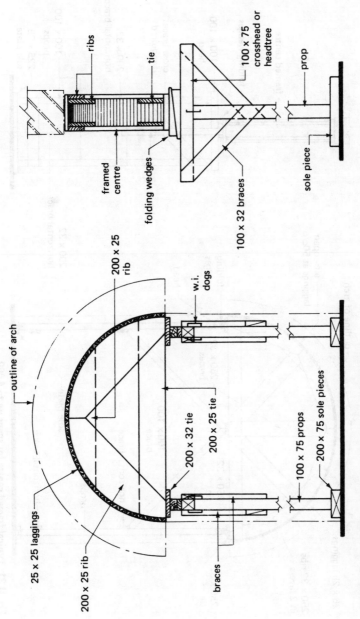

ribs

tie

100 × 75
crosshead or
headtree

prop

framed
centre

folding wedges

100 × 32 braces

sole piece

200 × 25
rib

w.i.
dogs

outline of arch

200 × 32 tie

200 × 25 tie

25 × 25 laggings

200 × 25 rib

braces

100 × 75 props

200 × 75 sole pieces

Fig. II.30 Typical framed centre for spans up to 1500 mm

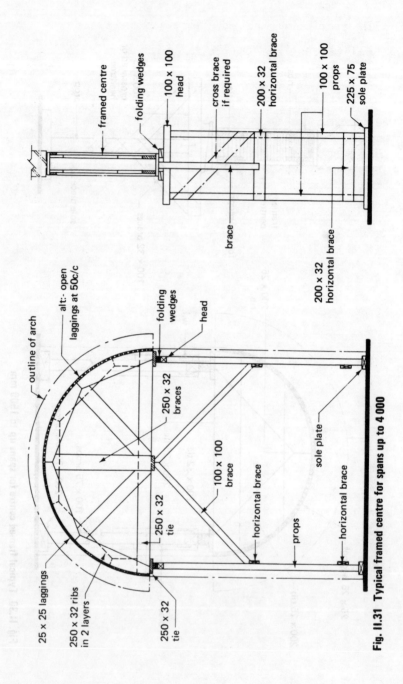

Fig. II.31 Typical framed centre for spans up to 4 000

framed centre

folding wedges

100 × 100 head

cross brace if required

200 × 32 horizontal brace

100 × 100 props

225 × 75 sole plate

brace

200 × 32 horizontal brace

outline of arch

alt:- open laggings at 50c/c

folding wedges

head

250 × 32 braces

100 × 100 brace

250 × 32 tie

25 × 25 laggings

250 × 32 ribs in 2 layers

250 × 32 tie

horizontal brace

props

horizontal brace

sole plate

The type of centre to be used will depend upon:

1. The weight to be supported.
2. The span.
3. The width of the soffit.

Generally soffits not wider than 150 mm will require one rib at least 50 mm wide and are usually called turning pieces. Soffits from 150-350 mm require two ribs which are framed together using horizontal tie members called laggings. Soffits over 350 mm require three or more sets of ribs. The laggings are used to tie the framed ribs together and to provide a base upon which the arch can be built. Close laggings are those which are touching each other, forming a complete seating for a gauged arch, and open laggings spaced at twice the width of the laggings, centre to centre, are used for rough arches.

If the arch is composed of different materials, for example, a stone arch with a relieving arch of brickwork, a separate centre for each material should be used. Typical examples of centres for brick arches are shown in Figs. II.29, II.30 and II.31.

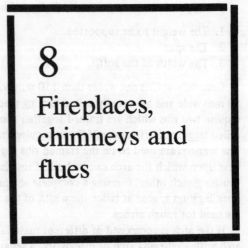

8
Fireplaces, chimneys and flues

The traditional method of providing heating in a domestic building is the open fire burning coal and/or wood but because of its low efficiency compared to modern heating appliances its use as the primary source of heating is declining. It must be noted that all combustible fuels, such as coal, smokeless fuels, oils, gas and wood, require some means of conveying the products of combustion away from the appliance or fireplace to the open air.

Terminology

Fireplace: this is the area in which combustion of the fuel takes place. It may be in the form of an open space with a fret or grate in which wood or coal is burnt, or a free standing appliance such as a slow combustion stove, room heater, oil burning appliance, gas fire or gas boiler.

Chimney: the fabric surrounding the flue and providing it with the necessary strength and protection.

Flue: this is strictly a void through which the products of combustion pass. It is formed by lining the inside of a chimney with a suitable lining material to give protection to the chimney fabric from the products of combustion, as well as forming a flue of the correct shape and size to suit the type of fuel and appliance being used.

Building Regulations J1 to J3 are concerned with heat-producing appliances and requires that they are installed with an adequate air supply for efficient working, provided with a suitable means of discharging the products of combustion to the outside air and installed so as to reduce the risk of the building catching fire. Approved Document J gives practical guidance to satisfy these regulations and makes particular reference to solid-fuel and oil-burning appliances with a rated output of up to 45 kW and it is this class with which a basic technology course is concerned.

When a fuel is burnt to provide heat it must have an adequate air supply to provide the necessary oxygen for combustion to take place. The air supply is drawn through the firebed in the case of solid fuels and injected in the case of oil burning appliances. A secondary supply of air is also drawn over the flames in both cases. In rooms where an open fire is used the removal of air induces ventilation and this helps to combat condensation. The air used in combustion is drawn through the appliance by suction provided by the flue and this is affected by the temperature, pressure, volume, cross-sectional area and surface lining of the flue.

To improve the efficiency of a flue it is advantageous for the chimney to be situated on an internal wall since this will reduce the heat losses to the outside air, also the flue gas temperature will be maintained and this will, in turn, reduce the risk of condensation within the flue. The condensation could cause corrosion of the chimney fabric due to the sulphur compounds which are a by product of the processes of combustion.

Chimneys

A chimney should be built as vertical as is practicable to give maximum flue gas flow—if bends are required the angle should lie between 45 and 30°. If an appliance is connected to a chimney the flue pipe should be as short as possible and continuous with an access for cleaning included at the base of the main flue. The chimney should be terminated above the roof so that it complies with the Building Regulations and is unaffected by adverse pressures which occur generally on the windward side of a roof and may cause down draughts. The use of a chimney terminal or pot could be to transfer the rectangular flue section to a circular cross-section giving better flue flow properties or to provide a cover to protect the flue from the entry of rain or birds. A terminal should be fixed so that it does not impede in any way the flow of flue gasses.

BUILDING REGULATIONS 1985

Heat-producing appliances are covered by Part J supported by Approved Document J which has four sections covering regulations J1 to J3. The main recommendations for appliances with a rated output of up to 45 kW together with typical construction details are shown in Figs II.32–II.35.

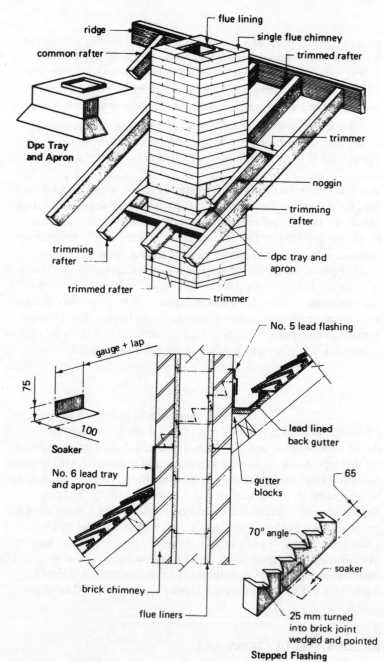

ridge

common rafter

flue lining

single flue chimney

trimmed rafter

trimmer

Dpc Tray
and Apron

noggin

trimming
rafter

trimming
rafter

dpc tray and
apron

trimmed rafter

trimmer

No. 5 lead flashing

gauge + lap

75

100

Soaker

lead lined
back gutter

No. 6 lead tray
and apron

gutter
blocks

65

70° angle

brick chimney

flue liners

soaker

25 mm turned
into brick joint
wedged and pointed

Stepped Flashing

Fig. II.32 Chimney construction

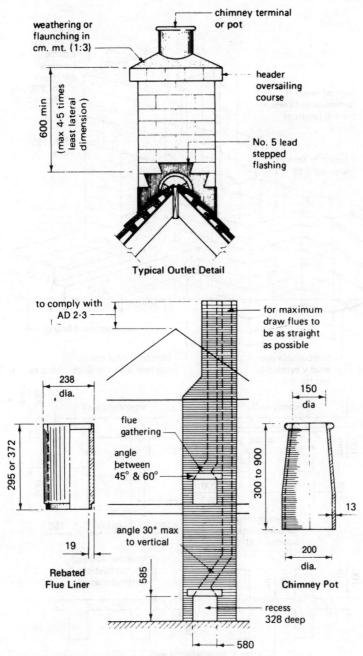

weathering or flaunching in cm. mt. (1:3)

chimney terminal or pot

header oversailing course

No. 5 lead stepped flashing

600 min (max 4·5 times least lateral dimension)

Typical Outlet Detail

to comply with AD 2·3

for maximum draw flues to be as straight as possible

238 dia.

295 or 372

150 dia

flue gathering

angle between 45° & 60°

300 to 900

13

19

angle 30° max to vertical

200 dia.

Rebated Flue Liner

585

Chimney Pot

recess 328 deep

580

Fig. II.33 Typical flue construction

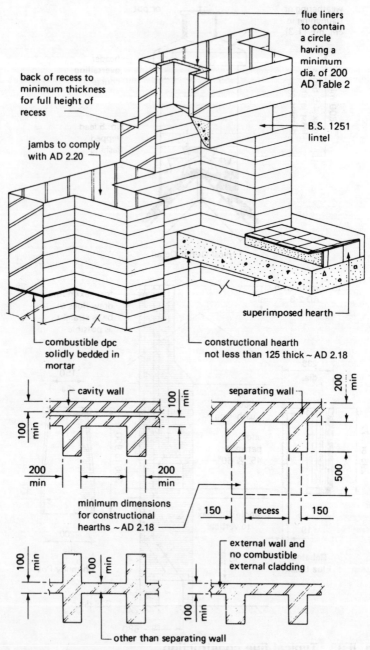

flue liners to contain a circle having a minimum dia. of 200 AD Table 2

back of recess to minimum thickness for full height of recess

jambs to comply with AD 2.20

B.S. 1251 lintel

combustible dpc solidly bedded in mortar

superimposed hearth

constructional hearth not less than 125 thick ~ AD 2.18

cavity wall
100 min
100 min
200 min
200 min

separating wall
200 min
500
150   recess   150

minimum dimensions for constructional hearths ~ AD 2.18

100 min
100 min

other than separating wall

external wall and no combustible external cladding
100 min

Fig. II.34   Fireplaces and Approved Document J

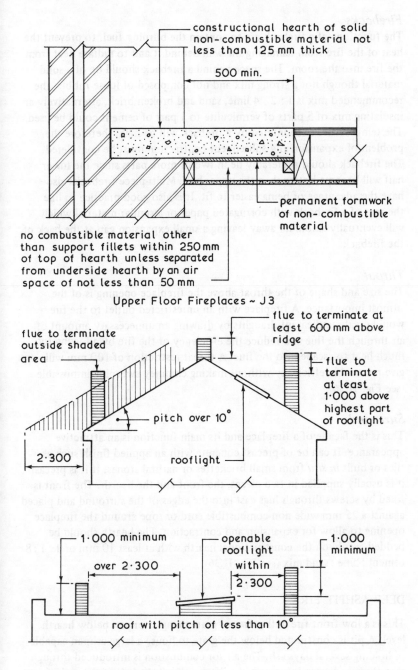

constructional hearth of solid non-combustible material not less than 125 mm thick

500 min.

permanent formwork of non-combustible material

no combustible material other than support fillets within 250 mm of top of hearth unless separated from underside hearth by an air space of not less than 50 mm

Upper Floor Fireplaces ~ J3

flue to terminate outside shaded area

flue to terminate at least 600 mm above ridge

flue to terminate at least 1·000 above highest part of rooflight

pitch over 10°

rooflight

2·300

1·000 minimum

1·000 minimum

over 2·300

openable rooflight within 2·300

roof with pitch of less than 10°

Fig. II.35 Fireplaces and outlets — Approved Document J

89

Firebacks

The functions of a fireback are to contain the burning fuel, to prevent the heat of the fire from damaging the wall behind it and to radiate heat from the fire into the room. The space behind a fireback should be of a solid material though not a strong mix and not composed of loose rubble, the recommended mix is 1 : 2 : 4 lime, sand and broken brick, alternatively an insulating mix of 5 parts of vermiculite to 1 part of cement could be used. The temperature encountered by a fireback will be high, therefore the problem of expansion and subsequent contraction must be considered. The fireback should preferably be in two or more parts since the lower half will become hotter than the upper half. Multi-piece firebacks also have the advantage of being easier to fit. It is also good practice to line the rear of a fireback with corrugated paper or a similar material which will eventually smoulder away leaving a small expansion gap at the back of the fireback.

Throat

The size and shape of the throat above the fireplace opening is of the utmost importance. A fireplace with an unrestricted outlet to the fire would create unpleasant draughts by drawing an unnecessary amount of air through the flue and reduce the efficiency of the fire by allowing too much heat to escape into the flue. A throat restriction of 100 mm will give reasonable efficiency without making chimney sweeping impossible (see Fig. II.36).

Surrounds

This is the façade of a fireplace and its main function is an attractive appearance. It can be of precast concrete with an applied finish such as tiles or built *in situ* from small brickettes or natural stones. If it is precast it is usually supplied in two pieces, the front and the hearth. The front is fixed by screws through lugs cast into the edges of the surround and placed against a 25 mm wide non-combustible cord or rope around the fireplace opening to allow for expansion and contraction. The hearth should be bedded evenly on the constructional hearth with at least 10 mm of 1 : 1 : 8 cement : lime : sand mix (see Fig. II.36).

DEEP ASHPIT FIRE

This is a low front fire with a fret or grate being at or just below hearth level. A pit is constructed below the grate to house a large ashpan capable of holding several days ash. The air for combustion is introduced through the ashpit and grate by means of a 75 mm diameter duct at the end of which is an air control regulator. If the floor is of suspended construction the air duct would be terminated after passing through the fender wall, whereas with a solid floor the duct must pass under the floor and be terminated beyond the external wall, the end being suitably guarded against the entry of vermin.

BACK BOILER

This an open fire of conventional design with part of the fireback replaced by a small boiler with a boiler flue and a control damper. It is used to supply hot water primarily for domestic use.

OPEN FIRE CONVECTORS

These are designed to increase the efficiency of an open fire by passing back into the room warm air as well as the radiated heat from the burning fuel. The open convector is a self-contained unit consisting of a cast-iron box containing the grate and fireback forming a convection chamber. The air in the chamber heats up and flows into the room by convection currents moving the air up and out of the opening at the top of the unit.

ROOM HEATERS

These are similar appliances to open fire convectors but they are designed to burn smokeless fuels and operate as a closed unit. In some models the strip glass front is, in fact, a door and can therefore be opened to operate as open fire. Room heaters are either freestanding, that is, fixed in front of the surround with a plate reducing the size of the fireplace opening to the required flue pipe size, or inset into the fireplace recess where the chimney flue aperture is reduced by a plate to receive the flue pipe of the appliance.

INDEPENDENT BOILERS

The function of this appliance is to heat water whether it is for a central heating system or merely to provide hot water for domestic use. It is generally fixed in the kitchen because of the convenience to the plumbing pipe runs and to utilise the background heat emitted. It must discharge into its own flue since it is a slow combustion appliance and the fumes emitted could, if joined to a common flue, cause a health hazard in other rooms of the building. Adequate access must be made for sweeping the flue and removing the fly ash that will accumulate with a smokeless fuel (see Fig. II.36).

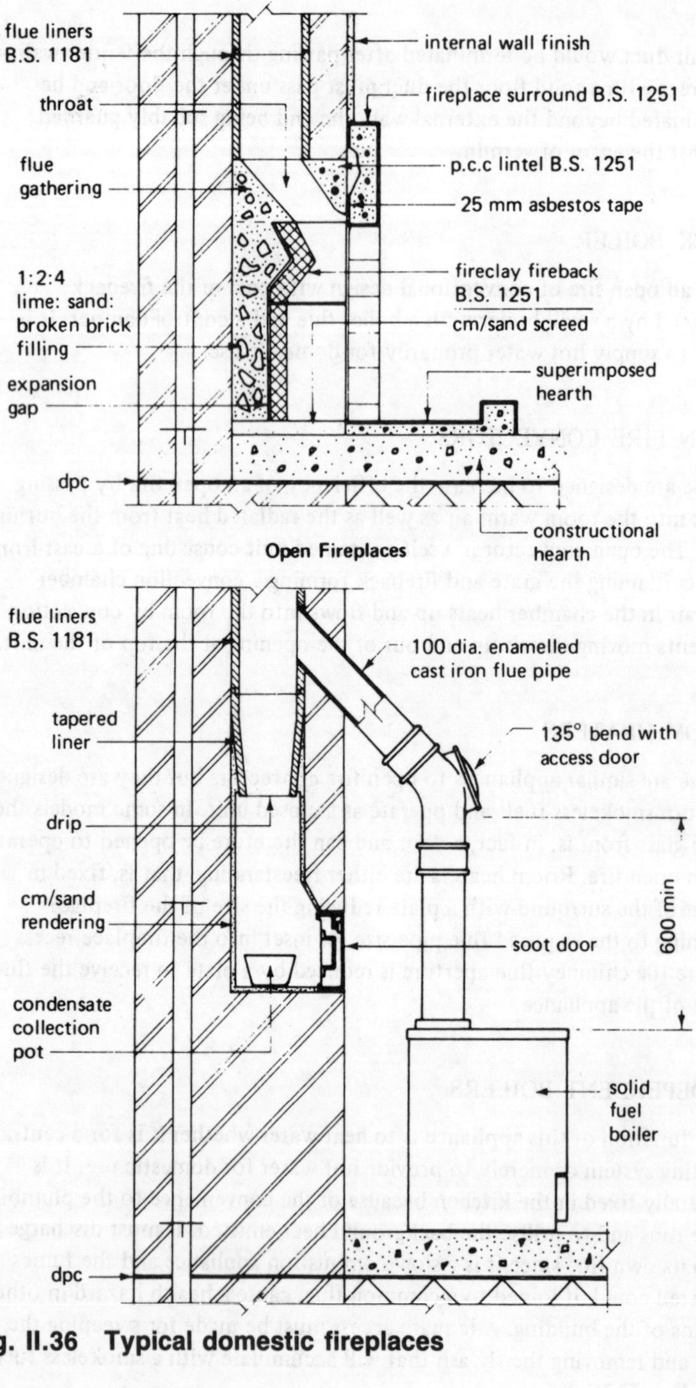
flue liners
B.S. 1181
internal wall finish
throat
fireplace surround B.S. 1251
flue
gathering
p.c.c lintel B.S. 1251
25 mm asbestos tape
1:2:4
lime: sand:
broken brick
filling
fireclay fireback
B.S. 1251
cm/sand screed
expansion
gap
superimposed
hearth
dpc
constructional
hearth
Open Fireplaces
flue liners
B.S. 1181
100 dia. enamelled
cast iron flue pipe
tapered
liner
135° bend with
access door
drip
cm/sand
rendering
soot door
600 min
condensate
collection
pot
solid
fuel
boiler
dpc

Fig. II.36 Typical domestic fireplaces

9
Simple framed buildings

The purpose of any framed building is to transfer the loads of the structure plus any imposed loads through the members of the frame to a suitable foundation. This form of construction can be clad externally with lightweight non-load bearing walls to provide the necessary protection from the elements and to give the required degree of comfort in terms of sound and thermal insulation. Framed buildings are particularly suitable for medium and high rise structures and for industrialised low rise buildings such as single storey factory buildings.

Frames can be considered under three headings:

Plane frames: fabricated in a flat plane and are usually called trusses or girders according to their elevation shape. They are designed as a series of connected rigid triangles which gives a lightweight structural member using the minimum amount of material; main uses are in roof construction and long span beams of light loading.

Space frames: similar in conception to a plane frame but are designed to span in two directions as opposed to the one-direction spanning of the plane frame. A variation of the space frame is the space deck which consists of a series of linked pyramid frames forming a lightweight roof structure. For details of these forms of framing textbooks on advanced building technology should be consulted.

Skeleton frames: basically these are a series of rectangular frames placed at right angles to one another so that the loads are transmitted from member to member until they are transferred through the foundations to the subsoil. Skeleton frames can be economically constructed of concrete or steel or a combination of the two. Timber skeleton frames, although possible, are generally considered to be uneconomic in this form. The choice of material for a framed structure can be the result of a number of factors such as site conditions, economics, availability of labour and materials, time factor, statutory regulations, capital costs, maintenance costs and personal preference.

Functions of skeleton frame members

Main beams: span between columns and transfer the live and imposed loads placed upon them to the columns.

Secondary beams: span between and transfer their loadings to the main beams. Primary function is to reduce the spans of the floors or roof being supported by the frame.

Tie beams: internal beams spanning between columns at right angles to the direction of the main beams and have the same function as a main beam.

Edge beams: as tie beam but spanning between external columns.

Columns: vertical members which carry the loads transferred by the beams to the foundations.

Foundation: the base(s), to which the columns are connected and serve to transfer the loadings to a suitable load-bearing subsoil.

Floors: may or may not be an integral part of the frame; they provide the platform on which equipment can be placed and on which people can circulate. Besides transmitting these live loads to the supporting beams they may also be required to provide a specific fire resistance, together with a degree of sound and thermal insulation.

Roof: similar to floors but its main function is to provide a weather-resistant covering to the uppermost floor.

Walls: the envelope of the structure which provides the resistance to the elements, entry of daylight, natural ventilation, fire resistance, thermal insulation and sound insulation.

The three major materials used in the construction of skeleton frames, namely, reinforced concrete, precast concrete and structural steel work, are considered in detail in the following chapters.

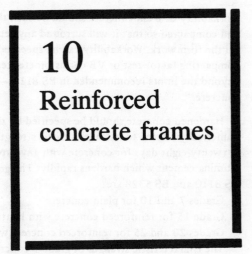

10
Reinforced concrete frames

Plain concrete is a mixture of cement, fine aggregate, coarse aggregate and water. Concrete sets to a rock-like mass due to a chemical reaction which takes place between the cement and water, resulting in a paste or matrix which binds the other constituents together. Concrete gradually increases its strength during the curing or hardening period to obtain its working strength in about twenty-eight days if ordinary Portland cement is used. The strength achieved will depend on a number of factors such as:

Type of cement used: in all cases the cement should conform to the appropriate British Standard.

Type and size of aggregates: in general aggregates should comply with the various British Standards and have nominal maximum coarse aggregate sizes of 40, 20, 14 and 10 mm.

Water: this should be clean, free from harmful matter and comply with the requirements of BS 3148.

Use of admixtures: individually admixtures can be used as accelerators, for air entrainment to give a weight reduction and added protection against water penetration, chemical and fungal attack. The instructions given by the manufacturer or engineer must be carefully followed since many admixtures if incorrectly used can have serious adverse effects on the hardened concrete.

Water/cement ratio: a certain amount of water is required to hydrate the cement and any extra water is needed only to produce workability. The workability of fresh concrete should be such that it can be handled, placed and compacted so that it will surround any reinforcement and completely fill the formwork. Workability can be specified in terms of a slump test, compacting factor test or VB consistometer test and should not vary beyond the limits recommended in BS 8110 — the structural use of concrete.

Hardened concrete should be specified by the grade required, which is numerically related to its characteristic strength or cube test strength taken at twenty-eight days for concrete with any type of cement except high alumina cement which hardens rapidly. The grades recommended in BS 8110 and BS 5328 are:

Grades 7 and 10 for plain concrete.

Grade 15 for reinforced concrete with lightweight aggregate.

Grades 20 and 25 for reinforced concrete with dense aggregates.

The characteristic strength of grade 7 concrete is 7.0 N/mm^2, and for grade 10 concrete it is 10.0 N/mm^2; similar values can be deduced for the other grades listed above.

Plain concrete in common with other brittle materials has a greater crushing or compressive strength than tensile strength. The actual ratio varies but plain concrete is generally considered to be ten times stronger in compression than in tension. If a plain concrete member is loaded so that tension is induced it will fail in tension when the compressive strength of concrete has only reached one-tenth of its ultimate value. If this weakness in tension can be reinforced in such a manner that the tensile resistance is raised to a similar value as its compressive strength the member will be able to support a load ten times that of plain concrete or alternatively for any given load a smaller section can be used if concrete is reinforced.

REINFORCEMENT

Any material specified for use as a reinforcement to concrete must fulfil certain requirements if an economic structural member is to be constructed. These basic requirements are:

1. Tensile strength.
2. Must be capable of achieving this tensile strength without undue strain.
3. Be of a material that can be easily bent to any required shape.
4. Its surface must be capable of developing an adequate bond between the concrete and the reinforcement to ensure that the required design tensile strength is obtained.
5. A similar coefficient of thermal expansion is required to prevent

unwanted stresses being developed within the member due to
temperature changes.
6. Availability at a reasonable cost which must be acceptable to the
overall design concept.

The material which meets all the above requirements is steel in the
form of bars, and is supplied in two basic types, namely mild steel and high
yield steel. Hot rolled steel bars are covered by BS 4449 which specifies a
characteristic strength of 250 N/mm^2 for mild steel and 410 N/mm^2 for
high yield steel. The surface of mild steel provides adequate bond but the
bond of high yield bars, being more critical with the higher stresses
developed, is generally increased by rolling on to the surface of the bar
longitudinal or transverse ribs. As an alternative to hot rolled steel bars,
cold worked steel bars complying with BS 4461 can be used. When bars are
cold worked they become harder, stiffer and develop a higher tensile
strength, this being 460 N/mm^2 for bars up to 16 mm nominal diameter
and 425 N/mm^2 for bars over 16 mm.

The range of diameters available for both round and deformed bars
recommended are 6, 8, 10, 12, 16, 20, 25, 32, and 40 with a recommended maximum length of 12.000 m. For pricing purposes the 16 and
20 mm bars are taken as basic with the diameters on either side becoming
more expensive as the size increases or decreases. A good design will limit
the range of diameters used together with the type of steel chosen to
achieve an economic structure and to ease the site processes of handling,
storage, buying and general confusion that can arise when the contractor is
faced with a wide variety of similar materials.

The bending of reinforcement can be carried out on site by using a
bending machine which shapes the cold bars by pulling them round a
mandrel. Small diameters can also be bent round simple jigs such as a
board with dowels fixed to give the required profile; large diameters may
need a power-assisted bending machine. Bars can also be supplied ready
bent and labelled so that only the fabrication processes takes place on site.

The bent reinforcement should be fabricated into cages for columns
and beams and into mats for slabs and walls. Where the bars cross or intersect one another they should be tied with soft iron wire, fixed with
special wire clips or tack welded to maintain their relative positions.
Structural members which require only small areas of reinforcement can be
reinforced with steel fabric which can be supplied in sheets or rolls.

Steel fabric for the reinforcement of concrete is covered by BS 4483,
which gives four basic preferred types. The fabric is factory-made by
welding or interweaving wires complying with the requirements of
BS 4482 to form sheets with a length of 4.800 m and a width of
2.400 m or alternatively rolls of 48.000 and 72.000 m in length with a

Square twisted bar

Plain round bar

Ribbed bar

Twisted ribbed bar

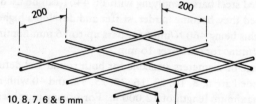

200

10, 8, 7, 6 & 5 mm
dia. ~ main and cross wires similar

Square mesh fabric ~ BS prefix 'A'

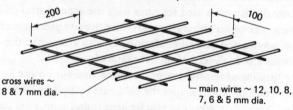

200

100

cross wires ~
8 & 7 mm dia.

main wires ~ 12, 10, 8,
7, 6 & 5 mm dia.

Structural mesh fabric ~ BS prefix 'B'

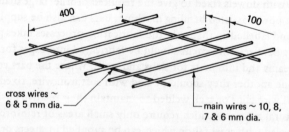

400

100

cross wires ~
6 & 5 mm dia.

main wires ~ 10, 8,
7 & 6 mm dia.

Long mesh fabric ~ BS prefix 'C'

Fig. II.37 Typical reinforcing bars and welded fabric

common width of 2.400 m. Each type has a letter prefix which is followed by a reference number which is the total cross sectional area of main bars in mm^2 per metre width. Typical examples of reinforcing bars and fabric are shown in Fig. II.37.

Before placing reinforcement into the formwork it should be brushed free of all loose rust and mill scale and be free of grease or oil, as the presence of any of these on the surface could reduce the bond and hence the strength of the reinforced concrete. Reinforcement must have a minimum cover of concrete to give the steel protection from corrosion due to contact with moisture and to give the structural member a certain degree of fire resistance. Nominal cover to reinforcement should always be equal to the size of the bar being used or where groups of bars are used at least the size of the largest diameter. BS 8110, Table 3.4 sets out the recommended nominal covers in relationship to the exposure condition and the concrete grade and Approved Document B sets out the minimum fire resistance for various purpose groups of buildings. It must be noted that the dimensions of structural members are also of importance so that failure of the concrete due to the high temperatures encountered during a fire is avoided before the reinforcement reaches its critical temperature. Table A3 of Approved Document B and Table 3.5 and Figure 3.2 in BS 8110 give suitable dimensions for various fire-resistance periods and methods of construction of reinforced concrete members.

To maintain the right amount of concrete cover during construction small blocks of concrete may be placed between the reinforcement and the formwork; alternatively, plastic clips or spacer rings can be used. Where top reinforcement such as in a slab has to be retained in position cradles or chairs made from reinforcing bar may have to be used. All forms of spacers must be of a material which will not lead to corrosion of the reinforcement or cause spalling of the hardened concrete.

Design

The design of reinforced concrete is the prerogative of the structural engineer and is a subject for special study but the technologist should have an understanding of the principles involved. The designer can by assessing the possible dead and live loads on a structural member calculate the reactions and effects such loadings will have on the member. This will enable him to determine where reinforcement is required and how much is needed. He bases his calculations upon the recommendations contained in BS 8110 which is for the structural use of concrete, together with formulae to enable him to determine bending moments, shear forces and the area of steel required. Typical examples of these forces for simple situations are shown in Fig. II.38.

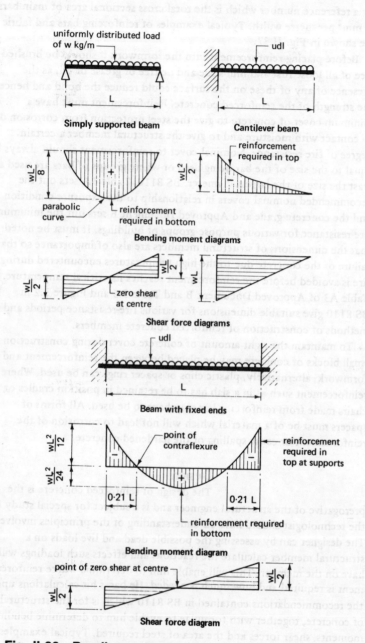

uniformly distributed load of w kg/m

L

Simply supported beam

udl

L

Cantilever beam

$\frac{wL^2}{8}$

parabolic curve

reinforcement required in bottom

$\frac{wL^2}{2}$

reinforcement required in top

Bending moment diagrams

$\frac{wL}{2}$

$\frac{wL}{2}$

zero shear at centre

$\frac{wL}{2}$

wL

Shear force diagrams

udl

L

Beam with fixed ends

point of contraflexure

$\frac{wL^2}{12}$

$\frac{wL^2}{24}$

reinforcement required in top at supports

reinforcement required in bottom

0.21 L

0.21 L

Bending moment diagram

point of zero shear at centre

$\frac{wL}{2}$

$\frac{wL}{2}$

Shear force diagram

Fig. II.38   Bending moment and shear force diagrams

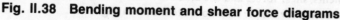

Reinforcement schedules and details

Once the engineer has determined the reinforcement required detail drawings can be prepared to give the contractor the information required to construct the structure. The drawings should give the following information:

1. Sufficient cross reference to identify the member in relationship to the whole structure.
2. All the necessary dimensions for design and fabrication of formwork.
3. Details of the reinforcement.
4. Minimum cover of concrete over reinforcement.
5. Concrete grade required if not already covered in the specification.

Reinforced concrete details should be prepared so that there is a distinct definition between the lines representing the outline of the member and those representing the reinforcement. Bars of a common diameter and shape are normally grouped together with the same reference number when included in the same member. To simplify the reading of reinforced concrete details it is common practice to show only one bar of each group in full together with the last bar position (see Fig. II.39).

The bars are normally bent and scheduled in accordance with the recommendations of BS 4466 which gives details of the common bending shapes, the method of setting out the bending dimensions, the method of calculating the total length of the bar required together with a shape code for use with data processing routines. A preferred form of bar schedule is also given, which has been designed to give easy cross reference to the detail drawing.

Reinforcement on detail drawings is annotated by a coding system to simplify preparation and reading of the details, for example:

9 R 1201 − 300 which can be translated as:

> 9 = total number of bars in the group
> 12 = diameter in mm
> 01 = bar mark number
> 300 = spacing centre to centre
> R = mild steel round bar

The code letter R could be replaced by code letter Y which is used for high yield round bars and high yield square twisted bars; other types of bars are coded with letter X. A typical reinforced concrete beam detail and schedule is shown in Fig. II.39. All other reinforced concrete details shown in this volume are intended to show patterns of reinforcement rather than detailing practice and are therefore shown in full without reference to bar diameters and types.

101

HOOKS, BENDS AND LAPS

To prevent bond failure bars should be extended beyond the section where there is no stress in the bar. The length of bar required will depend upon such factors as grade of concrete, whether the bar is in tension or compression and if the bar is deformed or plain. Hooks and bends can be used to reduce this anchorage length at the ends of bars and should be formed in accordance with the recommendations of BS 4466 (see Fig. II.40).

Where a transfer of stress is required at the end of a bar the bars may be welded or lapped.

BS 8110 recommends that laps and joints should only be made by the methods specified and at the positions shown on the drawings and as agreed by the engineer.

REINFORCED CONCRETE BEAMS

Beams can vary in their complexity of design and reinforcement from the very simple beam formed over an isolated opening such as those shown in Figs. II.39 and II.41 to the more common form encountered in frames where the beams transfer their loadings to the columns (see Fig. II.42).

When tension is induced into a beam the fibres will lengthen until the ultimate tensile strength is reached, when cracking and subsequent failure will occur. With a uniformly distributed load the position and value of tensile stress can easily be calculated by the structural engineer, but the problem becomes more complex when heavy point loads are encountered and this latter situation is considered beyond the scope of a second year course.

The correct design of a reinforced concrete beam will ensure that it has sufficient strength to resist both the compression and tensile forces encountered in the outer fibres, but it can still fail in the 'web' connecting the compression and tension areas. This form of failure is called shear failure and is in fact diagonal tension. Concrete has a limited amount of resistance to shear failure and if this is exceeded reinforcement must be added to provide extra resistance. Shear occurs at or near the supports as a diagonal failure line at an angle of approximately 45° to the horizontal and sloping downwards towards the support. A useful fact to remember is that zero shear occurs at the point of maximum bending (see Fig. II.38).

Reinforcement to resist shearing force may be either stirrups or inclined bars, or both. The total shearing resistance is the sum of the shearing resistances of the inclined bars and the stirrups, calculated separately if both are provided. Inclined or bent up bars should be at 45° to the

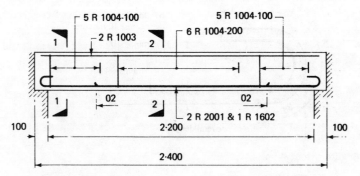

Elevation—beam 1-3 No. thus

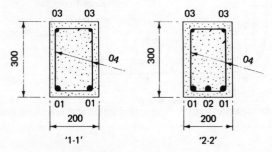

'1-1'   '2-2'

NB Cover to main bars 25 mm

Member	Bar mark	Type & size	No. of mbrs	No. in each	Total No.	Length of each bar†	Shape. All dimensions* are in accordance with BS 4466
Beam 1	1	R20	3	2	6	2660	⊂‾‾‾2300‾‾‾⊃
	2	R16	3	1	3	1400	straight
	3	R10	3	2	6	2300	straight
	4	R10	3	16	48	1000	250 ⌐☐ 150
† specified to nearest 25 mm					* specified to nearest 5 mm		

Fig. II.39 Typical R.C. beam details and schedule

103

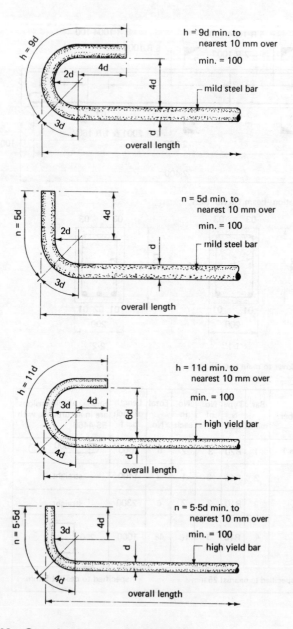

Fig. II.40 Standard hooks and bends

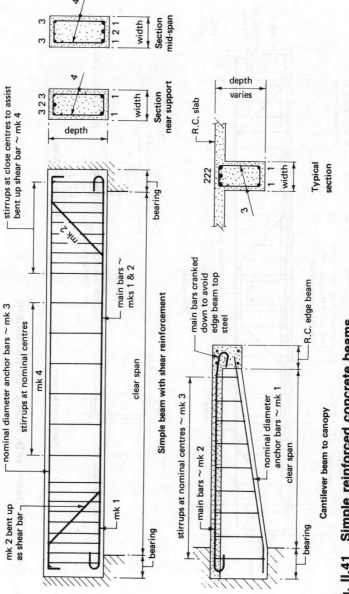

Section mid-span

Section near support

depth

width

Typical section

R.C. slab

depth varies

stirrups at close centres to assist bent up shear bar ~ mk 4

bearing

main bars ~ mks 1 & 2

mk 2

clear span

nominal diameter anchor bars ~ mk 3

stirrups at nominal centres mk 4

mk 2 bent up as shear bar

mk 1

bearing

Simple beam with shear reinforcement

main bars cranked down to avoid edge beam top steel

R.C. edge beam

stirrups at nominal centres ~ mk 3

main bars ~ mk 2

nominal diameter anchor bars ~ mk 1

clear span

bearing

Cantilever beam to canopy

Fig. II.41 Simple reinforced concrete beams

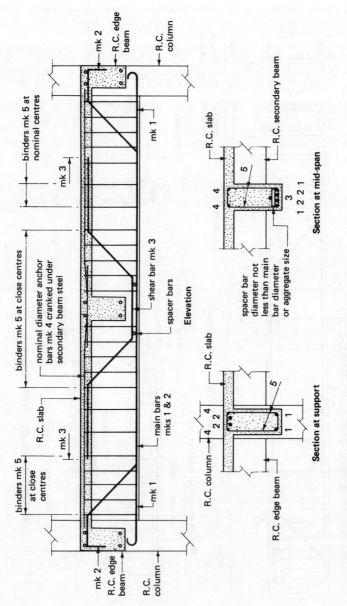

Fig. II.42 R.C. beam with heavy reinforcement

106

horizontal and positioned to cut the anticipated shear failure plane at right angles. These may be separate bars or alternatively main bars from the bottom of the beam which are no longer required to resist tension which can be bent up and carried over or onto the support to provide the shear resistance (see Figs. II.41 and II.42). Stirrups or binders are provided in beams, even where not required for shear resistance, to minimise shrinkage cracking and to form a cage for easy handling. The nominal spacing for stirrups must be such that the spacing dimension used is not greater than the lever arm of the section, which is the depth of the beam from the centre of the compression area to the centre of the tension area or 0.75 times the effective depth of the beam, which is measured from the top of the beam to the centre of the tension reinforcement. If stirrups are spaced at a greater distance than the lever arm it would be possible for a shearing plane to occur between consecutive stirrups, but if the centres of the stirrups are reduced locally about the position at which shear is likely to occur several stirrups may cut the shear plane and therefore the total area of steel crossing the shear plane is increased to offer the tensile resistance to the shearing force (see Figs. II.41 and II.42).

REINFORCED CONCRETE COLUMNS

A column is a vertical member carrying the beam and floor loadings to the foundation and is a compression member. Since concrete is strong in compression it may be concluded that provided the compressive strength of the concrete is not exceeded no reinforcement will be required. For this condition to be true the following conditions must exist:

1. Loading must be axial.
2. Column must be short, which can be defined as a column where the ratio of its effective height to its thickness does not exceed 12.
3. Cross section of the column must be large.

These conditions rarely occur in framed buildings, consequently bending is induced and the need for reinforcement to provide tensile strength is apparent. Bending in columns may be induced by one or more of the following conditions:

1. Load coupled with the slenderness of the column; a column is considered to be slender if the ratio of effective height to thickness exceeds 12.
2. Reaction to beams upon the columns, as the beam deflects it tends to pull the column towards itself thus inducing tension in the far face.

107

3. The reaction of the frame to wind loadings both positive and negative.

The minimum number of main bars in a column should not be less than four for rectangular columns and six for circular columns with a total cross section area of not less than 6% of the cross sectional area of the column and a minimum diameter of 12 mm. To prevent the slender main bars from buckling and hence causing spalling of the concrete, links or binders are used as a restraint. These should be at least one-quarter of the largest main bar diameter and at a pitch or spacing not greater than twelve times the main bar diameter. All bars in compression should be tied by a link passing around the bar in such a way that it tends to move the bar towards the centre of the column; typical arrangements are shown in Fig. II.43.

Where the junction between beams and columns occur there could be a clash of steel since bars from the beam may well be in the same plane as bars in the columns. To avoid this situation one group of bars must be bent or cranked into another plane; it is generally considered that the best practical solution is to crank the column bars to avoid the beam steel; typical examples of this situation together with a method using straight bars are shown in Fig. II.44. A similar situation can occur where beams of similar depth intersect; see cantilever beam example in Fig. II.41.

REINFORCED CONCRETE SLABS

A reinforced concrete slab will behave in exactly the same manner as a reinforced concrete beam and it is therefore designed in the same manner. The designer will analyse the loadings, bending moments, shear forces and reinforcement requirements on a slab strip 1.000 m wide. In practice the reinforcement will be fabricated to form a continuous mat. For light loadings a mat of welded fabric could be used.

There are three basic forms of reinforced concrete slabs, namely:
1. Flat slab floors or roofs.
2. Beam and slab floors or roofs.
3. Ribbed floors or roofs — see Part III.

Flat slabs

These are basically slabs contained between two plain surfaces and can be either simple or complex. The design of the complex form is based upon the slab acting as a plate in which the slab is divided into middle and column strips; the reinforcement being concentrated in the latter strips. For the purposes of a second year course only the simple flat slab will be dealt with in detail.

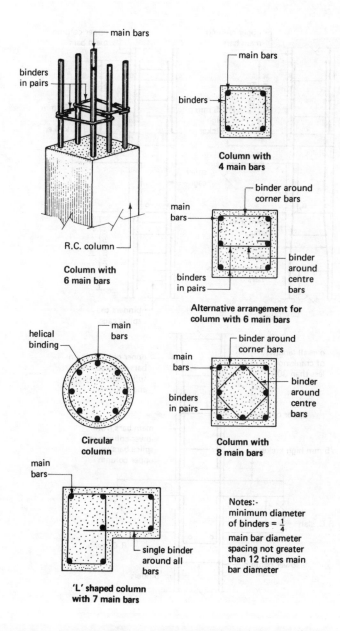

main bars

binders
in pairs

R.C. column

Column with
6 main bars

main bars

binders

Column with
4 main bars

binder around
corner bars

main
bars

binders
in pairs

binder
around
centre
bars

Alternative arrangement for
column with 6 main bars

helical
binding

main
bars

Circular
column

binder around
corner bars

main
bars

binders
in pairs

binder
around
centre
bars

Column with
8 main bars

main
bars

single binder
around all
bars

'L' shaped column
with 7 main bars

Notes:-
minimum diameter
of binders = $\frac{1}{4}$
main bar diameter
spacing not greater
than 12 times main
bar diameter

Fig. II.43 Typical R.C. column binding arrangements

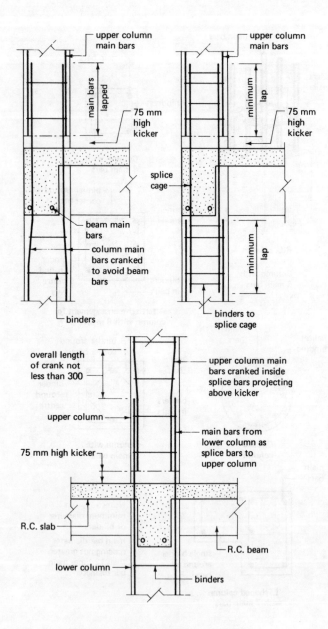

Fig. II.44   R.C. column and beam junctions

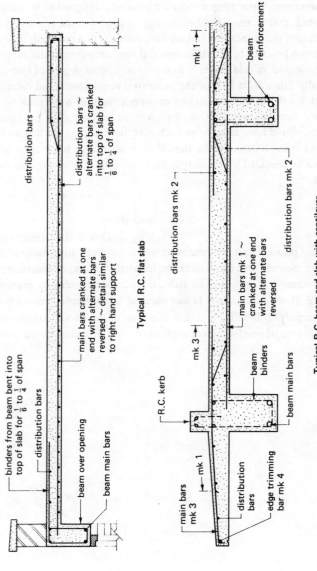

distribution bars

distribution bars ~
alternate bars cranked
into top of slab for $\frac{1}{6}$ to $\frac{1}{4}$ of span

main bars cranked at one
end with alternate bars
reversed ~ detail similar
to right hand support

binders from beam bent into
top of slab for $\frac{1}{6}$ to $\frac{1}{4}$ of span

distribution bars

beam over opening

beam main bars

Typical R.C. flat slab

mk 1

beam
reinforcement

distribution bars mk 2

distribution bars mk 2

main bars mk 1 ~
cranked at one end
with alternate bars
reversed

mk 3

R.C. kerb

beam
binders

beam main bars

main bars
mk 3

mk 1

distribution bars

edge trimming
bar mk 4

Typical R.C. beam and slab with cantilever

Fig. II.45 Typical R.C. slab details

111

Simple flat slabs can be thick and heavy but have the advantage of giving clear ceiling heights since there are no internal beams. They are generally economic up to spans of approximately 9.000 m and can be designed to span one way, that is across the shortest span, or to span in two directions. These simple slabs are generally designed to be simply supported, that is, there is no theoretical restraint at the edges and therefore tension is not induced and reinforcement is not required. However, it is common practice to provide some top reinforcement at the supports as anti-crack steel should there, in practice, be a small degree of restraint. Generally this steel is 50% of the main steel requirement and extends into the slab for 0.2 m of the span. An economic method is to crank up 50% of the main steel or every alternate bar over the support since the bending moment would have reduced to such a degree at this point it is no longer required in the bottom of the slab. If there is an edge beam the top steel can also be provided by extending the beam binders into the slab (see Fig. II.45.

Beam and slab

By adopting this method of design large spans are possible and the reinforcement is generally uncomplicated. A negative moment will occur over the internal supports necessitating top reinforcement; as with the flat slabs, this can be provided by cranked bars (see Fig. II.45). Each bar is in fact cranked but alternate bars are reversed thus simplifying bending and identification of the bars. Alternatively a separate mat of reinforcement supported on chairs can be used over the supports.

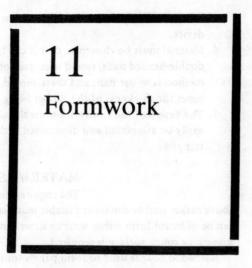

11
Formwork

Formworks for *in situ* concrete work may be described as a mould or box into which wet concrete can be poured and compacted so that it will flow and finally set to the inner profile of the box or mould. It is important to remember that the inner profile must be opposite to that required for the finished concrete so if, for example, a chamfer is required on the edge a triangle fillet must be inserted into the formwork.

To be successful in its function formwork must fulfil the following requirements:

1. It should be strong enough to support the load of wet concrete which is generally considered to be approximately 2 400 kg/m³.

2. It must not be able to deflect under load which would include the loading of wet concrete, self weight and any superimposed loads such as operatives and barrow runs over the formwork.

3. It must be accurately set out; concrete being a fluid when placed, it will take up the shape of the formwork which must therefore be of the correct shape, size and in the right position.

4. It must have grout-tight joints. Grout leakage can cause honey-combing of the surface or produce fins which have to be removed. The making good of defective concrete surfaces is both time consuming and costly. Grout leakage can be prevented by using sheet materials and sealing the joints with flexible foamed poly-urethane strip or by using a special self adhesive tape.

5. Form sizes should be designed so that they are the maximum size

which can easily be handled by hand or by a mechanical lifting device.

6. Material must be chosen so that it can be easily fixed using either double-headed nails, round wire nails or wood screws. The common method is to use nails and these should be at least two and a half times the thickness of the timber being nailed, in length.

7. The design of the formwork units should be such that they can easily be assembled and dismantled without any members being trapped.

MATERIALS

The requirements for formwork enumerated above makes timber the most suitable material for general formwork. It can be of board form either wrot or unwrot depending on whether a smooth or rough surface is required.

Softwood boards used to form panels for beam and column sides should be joined together by cross members over their backs at centres not exceeding twenty-four times the board's thickness.

The moisture content of the timber should be between 15 and 20% so that the moisture movement of the formwork is reduced to a minimum. If the timber is dry it will absorb moisture from the wet concrete which could weaken the resultant concrete member. It will also cause the formwork to swell and bulge which could give an unwanted profile to the finished concrete. If timber with a high moisture content is used it will shrink and cup, which could result in open joints and a leakage of grout.

Plywood is extensively used to construct formwork units since it is strong, light and supplied in sheets of 1.200 m wide with standard lengths of 2.400, 2.700 and 3.000 m. The quality selected should be an exterior grade and the thickness related to the anticipated pressures so that the minimum number of strengthening cleats on the back are required.

Chipboard can also be used as a formwork material but because of its lower strength will require more supports and stiffeners. The number of uses which can be obtained from chipboard forms is generally less than plywood, softwood boarding or steel.

Steel forms are generally based upon a manufacturer's patent system and within the constraints of that system are an excellent material. Steel is not so adaptable as timber but if treated with care will give thirty or forty uses, which is approximately double that of similar timber forms.

Mould oils and emulsions

Two defects which can occur on the surface of finished concrete are:

1. **Blow holes:** these are small holes being less than 15 mm in diameter caused by air being trapped between the formwork and the concrete face.
2. **Uneven colour:** this is caused by irregular absorption of water from the wet concrete by the formwork material. A mixture of old and new material very often accentuates this particular defect.

Mould oils can be applied to the inside surface of the formwork to alleviate these defects. To achieve a uniform colour an impervious material or lining is recommended but this will increase the risk of blow holes. Mould oils are designed to overcome this problem when using steel forms or linings by encouraging the trapped air to slide up the face of the formwork. A neat oil will encourage blow holes but will discourage uneven colour, whereas a mould oil incorporating an emulsifying agent will discourage blow holes and reduce uneven colouring. Great care must be taken when applying some mould oils since over oiling may cause some retardation of the setting of the cement. Emulsions are either drops of water in oil or conversely drops of oil in water and are easy to apply but should not be used in conjunction with steel forms since they encourage rusting. It should be noted that generally mould oils and emulsions also act as release agents and therefore it is essential that the oil or emulsion is only applied to the formwork and not to the reinforcement since this may cause a reduction of bond.

Formwork linings

To obtain smooth, patterned or textured surfaces the inside of a form can be lined with various materials such as oil-tempered hardboard, moulded rubber, moulded PVC and glass fibre reinforced polyester; the latter is also available as a complete form mould. When using any form of lining the manufacturer's instructions regarding sealing, fixing and the use of mould oils must be strictly followed to achieve a satisfactory result.

TYPES OF FORMWORK

Foundation formwork

Foundations to a framed building consist generally of a series of isolated bases or pads although if these pads are close together it may be more practicable to merge them together to form a strip. If the subsoil is firm and hard it may be possible to excavate the trench or pit for the foundations to the size and depth required and cast the concrete against the excavated faces. Where this method is not practicable formwork will be required. Side and end panels will be required and these should be firmly strutted against the excavation faces to resist the horizontal pressures of the wet concrete and to retain the formwork in the correct position. Ties will be required across the top of the form as a top restraint and these can be utilised to form the kicker for a reinforced concrete column or as a template for casting in the holding down bolts for precast concrete or structural steel columns (see Fig. II.46).

Column formwork

A column form or box consists of a vertical mould which has to resist considerable horizontal pressures in the early stages of casting. The column box should be located against a 75 mm-high plinth or kicker which has been cast monolithic with the base or floor. The kicker not only accurately positions the formwork but also prevents the loss of grout from the bottom edge of the form. The panels forming the column sides can be strengthened by using horizontal cleats or vertical studs which are sometimes called soldiers. The form can be constructed to the full storey height of the column with cut outs at the top to receive the incoming beam forms. The thickness of the sides does not generally provide sufficient bearing for the beam boxes and therefore the cut outs have a margin piece fixed around the opening to provide extra bearing (see Fig. II.47). It is general practice however to cast the columns up to the underside of the lowest beam soffit and to complete the top of the column at the same time as the beam using make-up pieces to complete the column and receive the beam intersections. The main advantage of casting full height columns is the lateral restraint provided by the beam forms, the disadvantage being the complexity of the formwork involved.

Column forms are held together with collars of timber or metal called yokes in the case of timber and clamps when made of metal. Timber yokes are purpose made whereas steel column clamps are adjustable within the limits of the blades (see Fig. II.48).

The spacing of the yokes and clamps should vary with the anticipated pressures, the greatest pressure occurring at the base of the column box. The actual pressure will vary according to:

1. Rate of placing.
2. Type of mix being used — generally the richer the mix the greater the pressure.
3. Method of placing — if vibrators are used pressures can increase up to 50% over hand placing and compacting.
4. Air temperature — the lower the temperature the slower is the hydration process and consequently higher pressures are encountered.

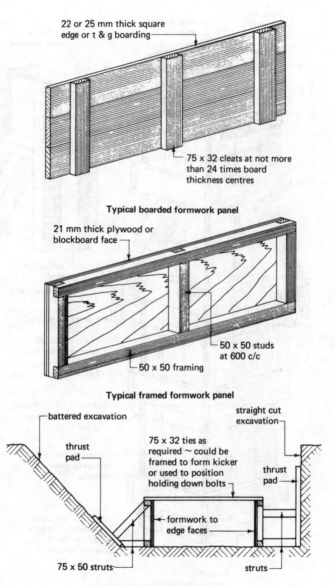

22 or 25 mm thick square edge or t & g boarding

75 x 32 cleats at not more than 24 times board thickness centres

Typical boarded formwork panel

21 mm thick plywood or blockboard face

50 x 50 studs at 600 c/c

50 x 50 framing

Typical framed formwork panel

battered excavation

straight cut excavation

thrust pad

75 x 32 ties as required ~ could be framed to form kicker or used to position holding down bolts

thrust pad

formwork to edge faces

75 x 50 struts

struts

Typical foundation formwork

Fig. II.46 Formwork to foundations

117

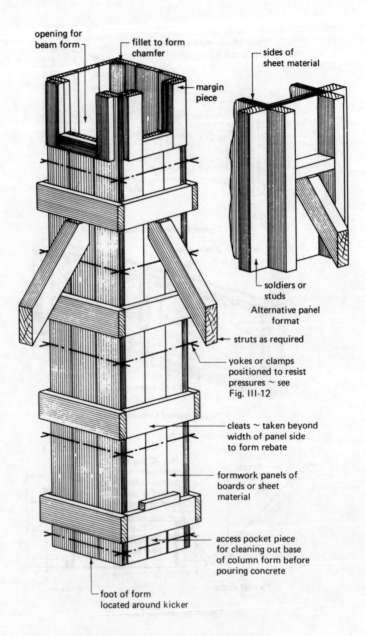

opening for beam form

fillet to form chamfer

margin piece

sides of sheet material

soldiers or studs

Alternative panel format

struts as required

yokes or clamps positioned to resist pressures ~ see Fig. III-12

cleats ~ taken beyond width of panel side to form rebate

formwork panels of boards or sheet material

access pocket piece for cleaning out base of column form before pouring concrete

foot of form located around kicker

Fig. II.47 Column formwork principles

118

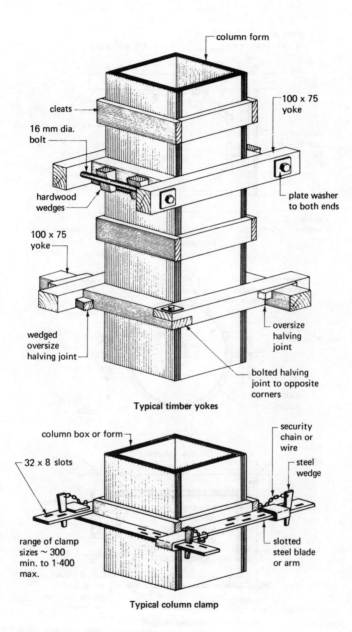

column form

cleats

16 mm dia.
bolt

hardwood
wedges

100 x 75
yoke

wedged
oversize
halving joint

100 x 75 yoke

plate washer
to both ends

oversize
halving
joint

bolted halving
joint to opposite
corners

Typical timber yokes

column box or form

32 x 8 slots

range of clamp
sizes ~ 300
min. to 1·400
max.

security
chain or
wire

steel wedge

slotted
steel blade
or arm

Typical column clamp

Fig. II.48   Typical column yokes and clamps

119

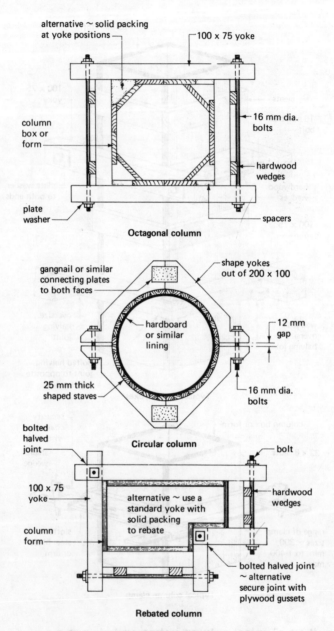

Fig. II.49 Shaped column forms and yokes

Some preliminary raking strutting is required to plumb and align the column forms in all situations. Free standing columns will need permanent strutting until the concrete has hardened but with tied columns the need for permanent strutting must be considered for each individual case.

Shaped columns will need special yoke arrangements unless they are being formed using a patent system. Typical examples of shaped column forms are shown in Fig. II.49.

Beam formwork

A beam form consists of a three-sided box which is supported by cross members called headtrees which are propped to the underside of the soffit board. In the case of framed buildings support to the beam box is also provided by the column form. The soffit board should be thicker than the beam sides since this member will carry the dead load until the beam has gained sufficient strength to be self supporting. Soffit boards should be fixed inside the beam sides so that the latter can be removed at an early date, this will enable a flow of air to pass around the new concrete and speed up the hardening process and also releasing the formwork for reuse at the earliest possible time. Generally the beam form is also used to support the slab formwork and the two structural members are then cast together. The main advantage of this method is that only one concrete operation is involved, although the complexity of the formwork is increased. If the beams and slabs are carried out as separate operations there is the possibility of a shear plane developing between the beam and floor slab; it would be advisable to consult the engineer before adopting this method of construction. Typical examples of beam forms are shown in Fig. II.50.

Structural steelwork has to be protected against corrosion and fire; one method is to encase the steel section with concrete. The steel frame is erected before casting the concrete encasement and in the case of beams it is possible to suspend the form box from the steel section by using a metal device called a hanger fixing or alternatively using a steel column clamp or timber yoke (see Fig. II.51). The hanger fixings are left embedded in the concrete encasing but the bolts and plate washers are recoverable for reuse. If only a haunch is cast around the bottom flange then the projecting hanger fixing wires can be cut off level with the concrete haunch or the floor units can be slotted to receive them.

Slab formwork

Floor or roof slab formwork is sometimes called shuttering and consists of panels of size that can be easily handled.

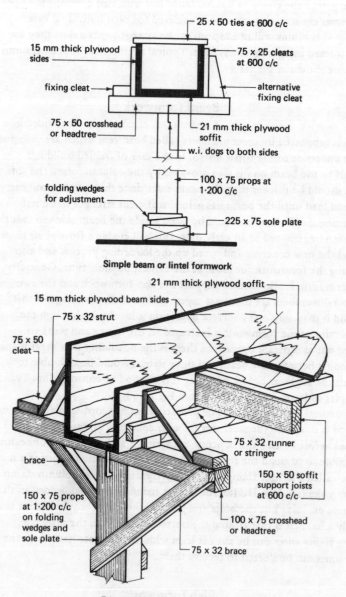

25 x 50 ties at 600 c/c

15 mm thick plywood sides

75 x 25 cleats at 600 c/c

fixing cleat

alternative fixing cleat

75 x 50 crosshead or headtree

21 mm thick plywood soffit

w.i. dogs to both sides

folding wedges for adjustment

100 x 75 props at 1·200 c/c

225 x 75 sole plate

Simple beam or lintel formwork

21 mm thick plywood soffit

15 mm thick plywood beam sides

75 x 32 strut

75 x 50 cleat

brace

150 x 75 props at 1·200 c/c on folding wedges and sole plate

75 x 32 runner or stringer

150 x 50 soffit support joists at 600 c/c

100 x 75 crosshead or headtree

75 x 32 brace

Edge beam and slab formwork

Fig. II.50 Typical beam formwork

The panels can be framed or joisted and supported by the beam forms with any intermediate propping which is required (see Fig. II.52). Adjustment for levelling purposes can be carried out by using small folding wedges between the joists or framing and the beam box.

SITEWORK

When the formwork has been fabricated and assembled the interior of the forms should be cleared of all rubbish, dirt and grease before the application of any mould oil or releasing agent. All joints and holes should be checked to ensure that they are grout tight.

The distance from the mixer to the formwork should be kept as short as possible to maintain the workability of the mix and to avoid as far as practicable double handling. Care must be taken when placing and compacting the concrete to ensure that the reinforcement is not displaced. The depth of concrete that can be placed in one lift will depend upon the mix and section size. If vibrators are used as the means of compaction this should be continuous during the placing of each batch of concrete until the air expulsion has ceased and care must be taken since over vibrating concrete can cause segregation of the mix.

The striking or removal of formwork should only take place upon instruction from the engineer or agent. The appropriate time at which it is safe to remove formwork can be assessed by tests on cubes taken from a similar batch mixed at the time the concrete was poured and cured under similar conditions. The characteristic cube strength should be 10 N/mm^2 or twice the stress to which the structure will then be submitted whichever is the greater before striking the formwork. If test cubes are not available the following table from BS 8110 can be used as a guide where ordinary Portland cement is used.

Location	Surface or air temperature	
	16° C	7° C
Vertical formwork	12 hours	18 hours
Slab soffits (Props left under)	4 days	6 days
Removal of Props	10 days	15 days
Beam soffits (Props left under)	10 days	15 days
Removal of Props	14 days	21 days

In very cold weather the above minimum periods should be doubled and when using rapid hardening Portland cement the above minimum periods can generally be halved.

Formwork must be removed slowly, as the sudden removal of the

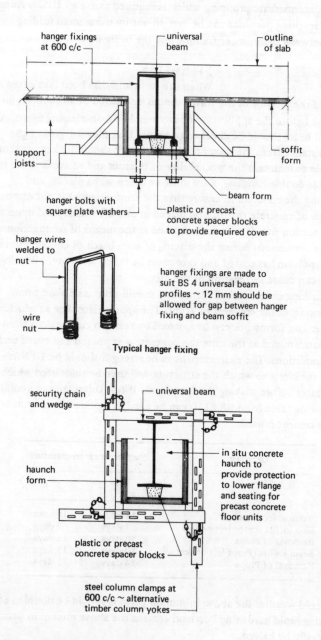

hanger fixings at 600 c/c

universal beam

outline of slab

support joists

soffit form

hanger bolts with square plate washers

plastic or precast concrete spacer blocks to provide required cover

beam form

hanger wires welded to nut

wire

nut

hanger fixings are made to suit BS 4 universal beam profiles ~ 12 mm should be allowed for gap between hanger fixing and beam soffit

Typical hanger fixing

security chain and wedge

universal beam

haunch form

in situ concrete haunch to provide protection to lower flange and seating for precast concrete floor units

plastic or precast concrete spacer blocks

steel column clamps at 600 c/c ~ alternative timber column yokes

Fig. II.51 Suspended formwork

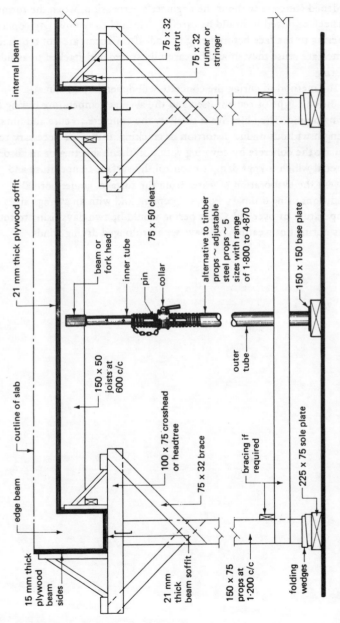

Labels in figure:

- internal beam
- 21 mm thick plywood soffit
- 75 × 32 strut
- 75 × 32 runner or stringer
- beam or fork head
- inner tube
- pin
- collar
- 75 × 50 cleat
- alternative to timber props ~ adjustable steel props ~ 5 sizes with range of 1·800 to 4·870
- 150 × 150 base plate
- 150 × 50 joists at 600 c/c
- 100 × 75 crosshead or headtree
- 75 × 32 brace
- outer tube
- bracing if required
- 225 × 75 sole plate
- outline of slab
- edge beam
- 15 mm thick plywood beam sides
- 21 mm thick beam soffit
- 150 × 75 props at 1·200 c/c
- folding wedges

Fig. II.52 Typical beam and slab formwork

125

wedges is equivalent to a shock load being placed upon the partly hardened concrete. Materials and/or plant should not be placed on the partly hardened concrete without the engineer's permission. When the formwork has been removed it should be carefully cleaned to remove any concrete adhering to the face before being reused. If the forms are not required for immediate reuse they should be carefully stored and stacked to avoid twisting.

The method of curing the concrete will depend upon climatic conditions, type of cement used and the average temperature during the curing period. The objective is to allow the concrete to cure and obtain its strength without undue distortion or cracking. It may be necessary to insulate the concrete by covering with polythene sheeting or an absorbent material which is kept damp to control the surface temperature and prevent the evaporation of water from the surface. Under normal conditions using ordinary Portland cement and with an average air temperature of over 10° C this period would be two days rising to four days during hot weather and days with prolonged drying winds.

12
Wall formwork

Formwork is a temporary mould into which wet concrete and reinforcement is placed to form a particular desired shape with a predetermined strength. Depending upon the complexity of the form, the relative cost of formwork to concrete can be as high as 75% of the total cost to produce the required member. A typical breakdown of percentage costs could be as follows:

Concrete	materials	28%	} 40%
	labour	12%	
Reinforcement	materials	18%	} 25%
	labour	7%	
Formwork	materials	15%	} 35%
	labour	20%	

The above breakdown shows that a building contractor will have to use an economic method of providing the necessary formwork if he is to be competitive in tendering since this is the factor over which he has most control.

The economic essentials of formwork can be listed thus:—

1. *Low cost* — only that amount of money necessary, which will produce the required form, to be expended.
2. *Strength* — careful selection of formwork materials and support members to obtain the most economic balance in terms of quantity used and continuing site activity around the assembled formwork.

3. *Finish* — selection of method, materials and if necessary linings to produce the desired result direct from the formwork. Applied finishes are usually specified and therefore only method is the real factor over which the builder would have any economic control.

4. *Assembly* — consideration must be given to the use of patent systems and mechanical handling plant.

5. *Material* — advantages of using either timber or steel should be considered; generally timber is lighter in weight and therefore larger units could be used, but steel will give more uses than timber although it cannot be repaired as easily.

6. *Design* — within the confines of the architectural and/or structural design formwork should be as repetitive and adaptable as possible.

A balance of the above essentials should be achieved, preferably at pretender stage, so that an economic and competitive cost can be calculated.

By the time the student has reached the first year of advanced construction technology he should have already studied the basic principles of formwork and in particular formwork for columns, beams and slabs. It is a useful exercise at this stage to recapitulate on these fundamentals (see Chapter 11) before proceeding with a study of wall formwork and patent systems.

In principle the design, fabrication and erection of wall formwork is similar to that already studied in the context of column formwork. Several basic methods are available which will enable a wall to be cast in large quantities, defined lifts or continuous from start to finish.

TRADITIONAL WALL FORMWORK
This usually consists of standard framed panels tied together over their backs with horizontal members called walings. The walings fulfil the same function as the yokes or column clamps of providing the resistance to the horizontal force of wet concrete. A 75 mm high concrete kicker is formed at the base of the proposed wall to enable the forms to be accurately positioned and to help prevent the loss of grout by seepage at the base of the form.

The usual assembly is to erect one side of the wall formwork ensuring that it is correctly aligned, plumbed and strutted. The steel reinforcement cage is inserted and positioned before the other side formwork is erected and fixed. Keeping the forms parallel and at the correct distance from one another is most important; this can be achieved by using precast concrete spacer blocks which are cast in, steel spacer tubes which are removed after casting and curing, the voids created being made good, or alternatively by using one of the many proprietary wall tie spacers available — see Fig. II.53.

To keep the number of ties required within acceptable limits horizontal members or walings are used, these also add to the overall rigidity of the formwork panels and help with alignment. Walings are best if composed of two members with a space between which will accommodate the shank of the wall tie bolt; this will give complete flexibility in positioning the ties and leave the waling timbers undamaged for eventual re-use. To ensure that the loads are evenly distributed over the pair of walings plate washers should be specified.

Plywood sheet is the common material used for wall formwork but this material is vulnerable to edge and corner damage. The usual format is therefore to make up wall forms as framed panels on a timber studwork principle with a plywood facing sheet screwed to the studs so that it can be easily removed and reversed to obtain the maximum number of uses.

Corners and attached piers need special consideration since the increased pressures at these points could cause the abutments between panels to open up, giving rise to unacceptable grout escape and a poor finish to the cast wall. The walings can be strengthened by including a loose tongue at the abutment position and extra bracing could be added to internal corners — see Fig. II.54. When considering formwork for attached piers it is usually necessary to have special panels to form the reveals.

CLIMBING FORMWORK

This is a method of casting a wall in set vertical lift heights using the same forms in a repetitive fashion thus obtaining maximum usage from a minimum number of forms. The basic formwork is as shown in Fig. II.55, which in the first lift is positioned against the kicker in the inverted position, the concrete is poured and allowed to cure after which the forms are removed, reversed and fixed to the newly cast concrete. After each casting and curing of concrete the forms are removed and raised to form the next lift until the required height has been reached.

It is possible to use this method for casting walls against an excavated or sheet piled back face using formwork to one side only by replacing the through wall tie spacers with loop wall ties — see Fig. II.55.

When using this single sided method adequate bracing will be required to maintain the correct wall thickness and when the formwork is reversed, after the first lift, it must be appreciated that the uprights or soldiers are acting as cantilevers and will therefore need to be stronger than those used in the double-sided version.

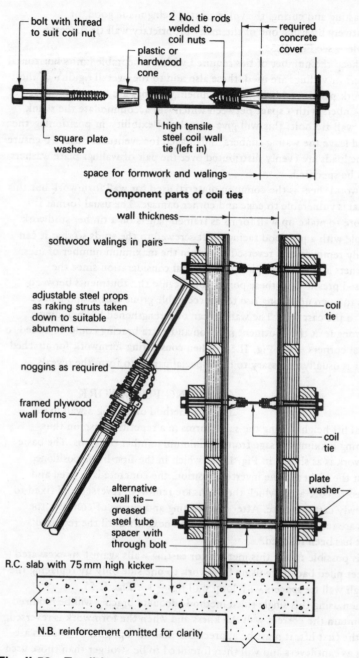

bolt with thread to suit coil nut

2 No. tie rods welded to coil nuts

required concrete cover

plastic or hardwood cone

square plate washer

high tensile steel coil wall tie (left in)

space for formwork and walings

Component parts of coil ties

wall thickness

softwood walings in pairs

adjustable steel props as raking struts taken down to suitable abutment

coil tie

noggins as required

framed plywood wall forms

coil tie

plate washer

alternative wall tie — greased steel tube spacer with through bolt

R.C. slab with 75 mm high kicker

N.B. reinforcement omitted for clarity

Fig. II.53 Traditional wall formwork — details 1

130

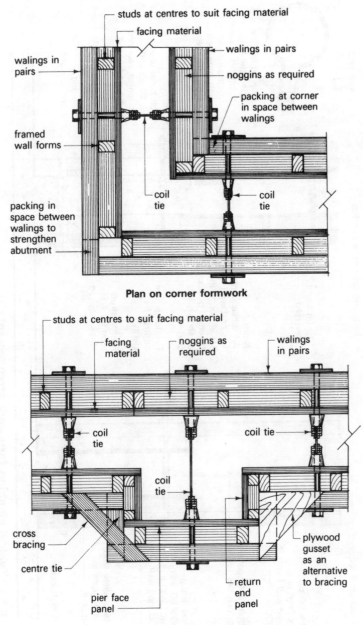

Plan on corner formwork

studs at centres to suit facing material

facing material

walings in pairs

walings in pairs

noggins as required

packing at corner in space between walings

framed wall forms

coil tie

coil tie

packing in space between walings to strengthen abutment

Plan on attached pier formwork

studs at centres to suit facing material

facing material

noggins as required

walings in pairs

coil tie

coil tie

coil tie

cross bracing

centre tie

pier face panel

return end panel

plywood gusset as an alternative to bracing

Fig. II.54 Traditional wall formwork — details 2

131

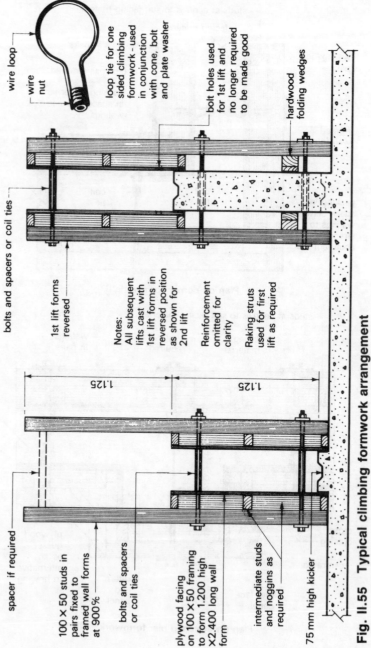

wire loop

wire nut

loop tie for one sided climbing formwork - used in conjunction with cone, bolt and plate washer

bolts and spacers or coil ties

bolt holes used for 1st lift and no longer required to be made good

1st lift forms reversed

hardwood folding wedges

Notes:
All subsequent lifts cast with 1st lift forms in reversed position as shown for 2nd lift

Reinforcement omitted for clarity

Raking struts used for first lift as required

1.125

1.125

spacer if required

100 × 50 studs in pairs fixed to framed wall forms at 900%

bolts and spacers or coil ties

plywood facing on 100 × 50 framing to form 1.200 high ×2.400 long wall form

intermediate studs and noggins as required

75 mm high kicker

Fig. II.55 Typical climbing formwork arrangement

132

SLIDING FORMWORK

This is a system of formwork which slides continuously up the face of the wall being cast by climbing up and being supported by a series of hydraulic jacks operating on jacking rods. The whole wall is therefore cast as a monolithic and jointless structure making the method suitable for structures such as water towers, chimneys and the cores of multi-storey buildings which have repetitive floors.

Since the system is a continuous operation good site planning and organisation is very essential and will involve the following aspects:

1. Round-the-clock working which will involve shift working and artificial lighting to enable work to proceed outside normal daylight hours.
2. Careful control of concrete supply to ensure that stoppages of the lifting operation are not encountered. This may mean having standby plant as an insurance against mechanical breakdowns.
3. Suitably trained staff accustomed to this method of constructing *in situ* concrete walls

The actual architectural and structural design must be suitable for the application of a slipform system; generally the main requirements are a wall of uniform thickness with a minimum number of openings and a height of at least 20.000 m to make the cost of equipment, labour and planning an economic proposition.

The basic components of slip formwork are:

1. *Side forms* — these need to be strongly braced and are loadbearing of timber and/or steel construction. Steel forms are heavier than timber, more difficult to assemble and repair but they have lower frictional loading, are easier to clean and have better durability. Timber forms are lighter, have better flexibility, easier to repair and are generally favoured. A typical timber form would consist of a series of 100 × 25 planed straight grained staves assembled with a 2 mm wide gap between consecutive boards to allow for swelling which could give rise to unacceptable friction as the forms rise. The forms are usually made to a height of 1.200 m with an overall sliding clearance of 6 mm by keeping the external panel plumb and the internal panel tapered so that it is 3 mm in at the top and 3 mm out at the bottom, giving the true wall thickness, in the centre position of the form. The side forms must be adequately stiffened with horizontal walings and vertical puncheons to resist the lateral pressure of concrete and transfer the loads of working platforms to the supporting yokes.

2. *Yokes* — assist in supporting the suspended working platforms and transfers the platform and side form loads to the jacking rods. Yokes are usually made of framed steelwork suitably braced and designed to provide the necessary bearings for the working platforms.
3. *Working platforms* — three working levels are usually provided, the first is situated above the yokes at a height of about 2.000 m above the top of the wall forms for the use of the steel fixers. The second level is a platform over the entire inner floor area at a level coinciding with the top of the wall forms and is used by the concrete gang, for storage of materials, to carry levelling instruments and jacking control equipment. It is worth noting that this decking could ultimately be used as the soffit formwork to the roof slab if required. The third platform is in the form of a hanging or suspended scaffold usually to both sides of the wall and is to give access to the exposed freshly cast concrete below the slip formwork for the purpose of finishing operations.
4. *Hydraulic jacks* — the jacks used are usually specified by their load bearing capacities such as 3 tonnes and 6 tonnes and consist of two clamps operated by a piston. The clamps operate on a jacking rod of 25 to 50 mm diameter according to the design load and are installed in banks operated from a central control to give an all round consistent rate of climb. The upper clamp grips the jacking rod and the lower clamp, being free, rises, pulling the yoke and platforms with it until the jack extension has been closed. The lower clamp now grips the climbing rod whilst the upper clamp is released and raised to a higher position when the lifting cycle is recommenced. Factors such as temperature and concrete quality affect the rate of climb but typical speeds are between 150 and 450 mm per hour.

The upper end of the jacking rod is usually encased in a tube or sleeve to overcome the problem of adhesion between the rod and the concrete. The jacking rod therefore remains loose in the cast wall and can be recovered at the end of the jacking operation. The 2.500 to 4.000 m lengths of rod are usually joined together with a screw joint arranged so that all such joints do not occur at the same level. A typical diagrammatic arrangement of sliding formwork is shown in Fig. II.56.

The site operations commence with the formation of a substantial kicker 300 mm high incorporating the wall and jacking rod starter bars. The wall forms are assembled and fixed together with the yokes, upper working platforms and jacking arrangement after which the initial concrete lift is poured. The commencing rate of climb must be slow to allow time for the first batch of concrete to reach a suitable condition before emerging from beneath the sliding formwork. A standard or planned rate

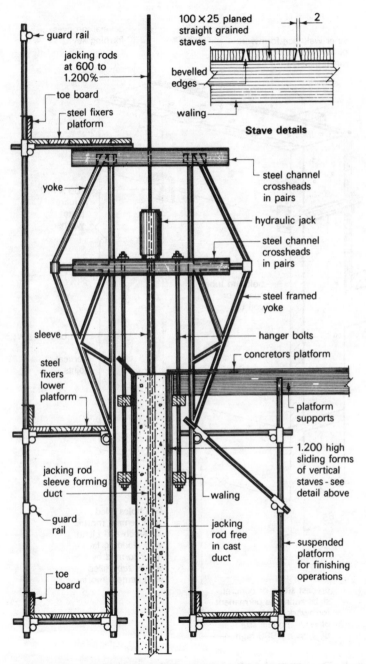

guard rail

jacking rods at 600 to 1.200%

toe board

steel fixers platform

yoke

sleeve

steel fixers lower platform

jacking rod sleeve forming duct

guard rail

toe board

100 × 25 planed straight grained staves

2

bevelled edges

waling

Stave details

steel channel crossheads in pairs

hydraulic jack

steel channel crossheads in pairs

steel framed yoke

hanger bolts

concretors platform

platform supports

1.200 high sliding forms of vertical staves - see detail above

waling

jacking rod free in cast duct

suspended platform for finishing operations

Fig. II.56   Diagrammatic arrangement of sliding formwork

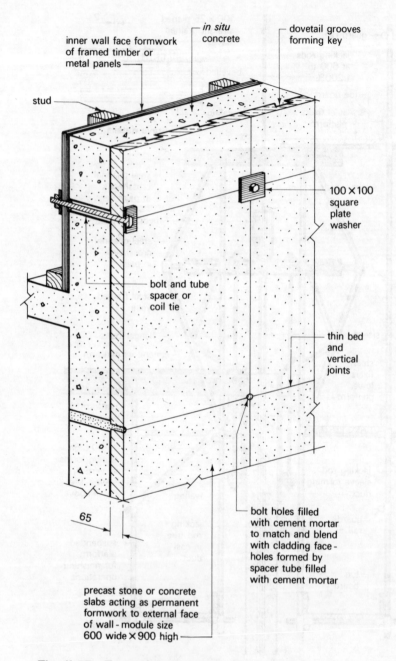

inner wall face formwork of framed timber or metal panels

stud

in situ concrete

dovetail grooves forming key

100 × 100 square plate washer

bolt and tube spacer or coil tie

thin bed and vertical joints

65

bolt holes filled with cement mortar to match and blend with cladding face - holes formed by spacer tube filled with cement mortar

precast stone or concrete slabs acting as permanent formwork to external face of wall - module size 600 wide × 900 high

Fig. II.57 Example of permanent formwork

of lift is usually reached within about 16 hours after commencing the lifting operation.

Openings can be formed in the wall by using framed formwork with splayed edges, to reduce friction, tied to the reinforcement. Small openings can be formed using blocks of expanded polystyrene which should be 75 mm less in width than the wall thickness so that a layer of concrete is always in contact with the sliding forms to eliminate friction. The concrete cover is later broken out and the blocks removed. Chases for floor slabs can be formed with horizontal boxes drilled to allow the continuity reinforcement to be passed through and to be bent back within the thickness of the wall so that when the floor slabs are eventually cast the reinforcement can be pulled out into its correct position.

PERMANENT FORMWORK

In certain circumstances formwork is left permanently in place because of the difficulty and/or cost of removing it once the concrete has been cast. A typical example of these circumstances would be when a beam and slab raft foundation with shallow upstand beams and an *in situ* slab have been constructed. Apart from the cost aspect, consideration must be given to any nuisance such an arrangement could cause in the finished structure, such as the likelihood of fungi or insect attack and the possible risk of fire.

Permanent formwork can also be a means of utilising the facing material as both formwork and outer cladding especially in the construction of *in situ* reinforced concrete walls. The external face or cladding is supported by the conventional internal face formwork, which can in certain circumstances overcome the external strutting or support problems often encountered with high rise structures.

This method is however generally limited to thin small modular facing materials, the size of which is governed by the supporting capacity of the internal formwork. Fig. II.57 shows a typical example of the application of this aspect of permanent formwork.

The methods described above for the construction of *in situ* reinforced concrete walls can also be carried out by using a patent system of formwork.

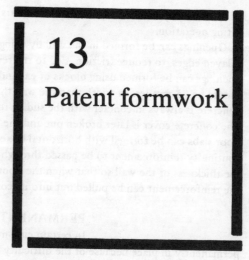

13
Patent formwork

Patent formwork is sometimes called system formwork and is usually identified by the manufacturer's name. All proprietary systems have the same common aim and most are similar in their general approach to solving some of the problems encountered with formwork for modular designed or repetitive structures. As shown in the previous chapter, formwork is one area where the contractor has most control over the method and materials to be used in forming an *in situ* reinforced concrete structure. In trying to design or formulate the ideal system for formwork the following must be considered:

1. *Strength* — to carry the concrete and working loads.
2. *Lightness without strength reduction* — to enable maximum size units to be employed.
3. *Durability without prohibitive costs* — to gain maximum usage of materials.
4. *Good and accurate finish straight from the formwork* — to reduce the costly labour element of making good and patching, which in itself is a difficult operation to accomplish without it being obvious that this kind of treatment was found necessary.
5. *Erection and dismantling times.*
6. *Ability to employ unskilled or semi-skilled labour to carry out the work.*

Patent or formwork systems have been devised to satisfy most of the above listed requirements by the standardisation of forms and by easy methods of securing and bracing the positioned formwork.

The major component of any formwork system is the unit panel which should fulfil the following requirements:

1. Available in a wide variety of sizes based on a standard module, usually multiples and submultiples of 300 mm.
2. Manufactured from durable materials.
3. Covered with a facing material which is durable and capable of producing the desired finish.
4. Units should be interchangeable so that they can be used for beams, columns and slabs.
5. Formed so that they can be easily connected together to form large unit panels.
6. Lightweight so that individual unit panels can be handled without mechanical aid.
7. Designed so that the whole formwork can be assembled and dismantled easily by unskilled or semi-skilled labour.
8. Capable of being adapted so that non-standard width inserts of traditional formwork materials can be included where lengths or widths are not exact multiples of the unit panels.

Most unit panels consist of a framed tray made from light metal angle or channel sections stiffened across the width as necessary. The edge framing is usually perforated with slots or holes to take the fixing connectors and waling clips or clamps. The facing is of sheet metal or plywood, some manufacturers offering a choice. Longitudinal stiffening and support is given by clamping, over the backs of the assembled panels, special walings of hollow section or in many systems standard scaffold tubes are used. Vertical support where required can be given by raking standard adjustable steel props, or where heavy loadings are encountered most systems have special vertical stiffening and support arrangements based on designed girder principles which also provide support for an access and working platform. Spacing between opposite forms is maintained using wall ties or similar devices as previously described for traditional formwork.

Walls which are curved in plan can be formed in a similar manner to the straight wall techniques described above using a modified tray which has no transverse stiffening members thus making it flexible. Climbing formwork can also be carried out using system formwork components but instead of reversing the forms as described for traditional formwork climbing shoes are bolted to the cast section of the wall to act as bearing corbels to supports the soldiers for each lift.

When forming beams and columns the unit panels are used as side or soffit forms held together with steel column clamps, or in the case of beams conventional strutting or alternatively using wall ties through the beam thickness. Some manufacturers produce special beam box clamping devices which consist of a cross member surmounted by attached and adjustable triangulated struts to support the side forms.

Many patent systems for the construction of floor slabs which require propping during casting use basic components of unit panels, narrow width (150 mm) filler panels, special drop head adjustable steel props, joists and standard scaffold tubes for bracing. The steel joists are lightweight, purpose made and are supported on the secondary head of the prop in the opposite direction to the filler panels which are also supported by the secondary or drop head of the prop. The unit panels are used to infill between the filler panels and pass over the support joist, the upper head of the prop being at the same level as the slab formwork and is indeed part of the slab soffit formwork.

After casting the slab and allowing it to gain sufficient strength the whole of the slab formwork can be lowered and removed leaving the undisturbed prop head to give the partially cured slab a degree of support. This method enables the formwork to be removed at a very early stage releasing the unit panels, filler pieces and joists for re-use and at the same time accelerating the drying out of the concrete by allowing a free air flow on both sides of the slab.

Slabs of moderate spans (up to 7.500 m) which are to be cast between load bearing walls or beams without the use of internal props can be formed using unit panels supported by steel telescopic floor centres. These centres are made in a simple lattice form extending in one or two directions according to span and are light enough to be handled by one man. They are precambered to compensate any deflections when loaded. Typical examples of system formwork are shown in Figs. II.58 to II.61.

TABLE FORMWORK

This special class of formwork has been devised for use when casting large repetitive floor slabs in medium to high rise structures. The main objective is to reduce the time factor in erecting, striking and re-erecting slab formwork by creating a system of formwork which can be struck as an entire unit, removed, hoisted and repositioned without any dismantling.

The basic requirements for a system of table formwork can be listed thus:

1. A means of adjustment for aligning and levelling the forms.

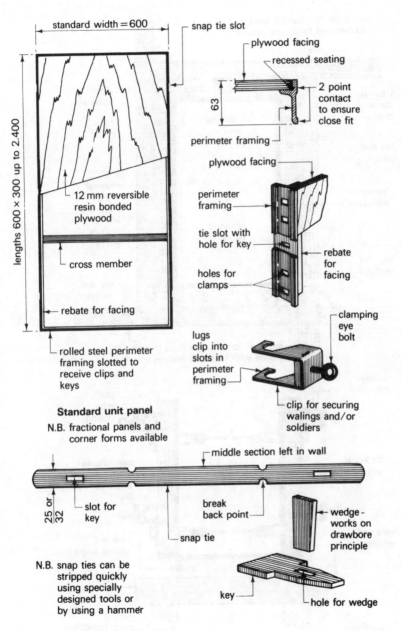

standard width = 600

snap tie slot

plywood facing

recessed seating

63

2 point contact to ensure close fit

perimeter framing

lengths 600 × 300 up to 2.400

plywood facing

perimeter framing

tie slot with hole for key

holes for clamps

rebate for facing

12 mm reversible resin bonded plywood

cross member

rebate for facing

clamping eye bolt

lugs clip into slots in perimeter framing

clip for securing walings and/or soldiers

rolled steel perimeter framing slotted to receive clips and keys

Standard unit panel

N.B. fractional panels and corner forms available

middle section left in wall

25 or 32

slot for key

break back point

snap tie

wedge - works on drawbore principle

key

hole for wedge

N.B. snap ties can be stripped quickly using specially designed tools or by using a hammer

for application details see Figs VI.7 to VI.9

Fig. II.58 System formwork — typical components

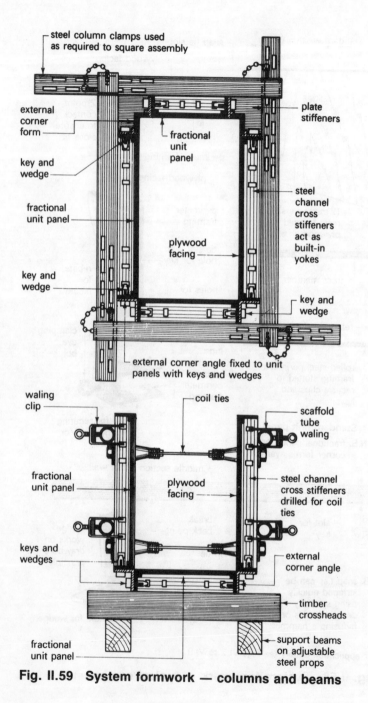

steel column clamps used as required to square assembly

external corner form

key and wedge

fractional unit panel

key and wedge

fractional unit panel

plate stiffeners

fractional unit panel

plywood facing

steel channel cross stiffeners act as built-in yokes

key and wedge

external corner angle fixed to unit panels with keys and wedges

waling clip

coil ties

scaffold tube waling

fractional unit panel

plywood facing

steel channel cross stiffeners drilled for coil ties

keys and wedges

external corner angle

timber crossheads

fractional unit panel

support beams on adjustable steel props

Fig. II.59 System formwork — columns and beams

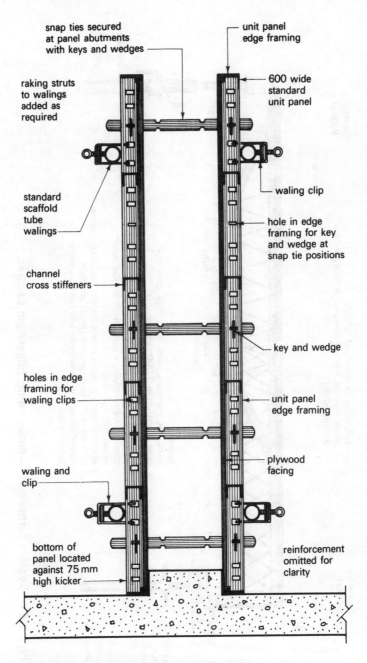

snap ties secured at panel abutments with keys and wedges

unit panel edge framing

raking struts to walings added as required

600 wide standard unit panel

waling clip

standard scaffold tube walings

hole in edge framing for key and wedge at snap tie positions

channel cross stiffeners

key and wedge

holes in edge framing for waling clips

unit panel edge framing

plywood facing

waling and clip

bottom of panel located against 75 mm high kicker

reinforcement omitted for clarity

Fig. II.60 System formwork — walls

144

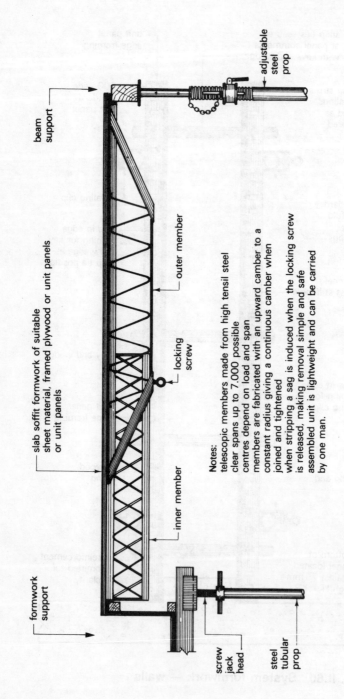

slab soffit formwork of suitable sheet material, framed plywood or unit panels

beam support

adjustable steel prop

outer member

locking screw

inner member

formwork support

Notes:

telescopic members made from high tensil steel clear spans up to 7.000 possible centres depend on load and span members are fabricated with an upward camber to a constant radius giving a continuous camber when joined and tightened when stripping a sag is induced when the locking screw is released, making removal simple and safe assembled unit is lightweight and safe and can be carried by one man.

screw jack head

steel tubular prop

Fig. II.61 System formwork — slabs using telescopic centres

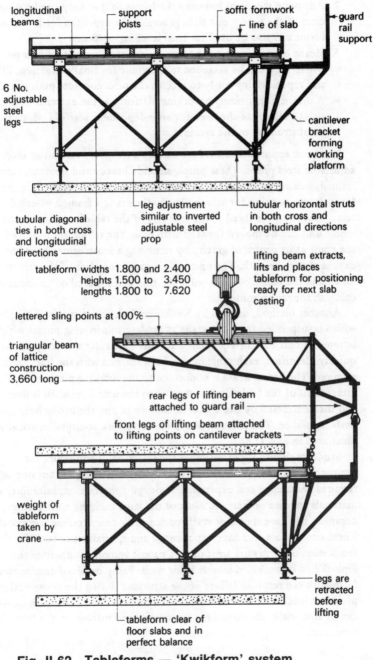

longitudinal beams — support joists — soffit formwork — line of slab — guard rail support

6 No. adjustable steel legs

cantilever bracket forming working platform

tubular diagonal ties in both cross and longitudinal directions

leg adjustment similar to inverted adjustable steel prop

tubular horizontal struts in both cross and longitudinal directions

tableform widths 1.800 and 2.400
heights 1.500 to 3.450
lengths 1.800 to 7.620

lifting beam extracts, lifts and places tableform for positioning ready for next slab casting

lettered sling points at 100%

triangular beam of lattice construction 3.660 long

rear legs of lifting beam attached to guard rail

front legs of lifting beam attached to lifting points on cantilever brackets

weight of tableform taken by crane

legs are retracted before lifting

tableform clear of floor slabs and in perfect balance

Fig. II.62 Tableforms — 'Kwikform' system

145

2. Adequate means of lowering the forms so that they can be dropped clear of the newly cast slab; generally the provision for lowering the forms can also be used for final levelling purposes.
3. Means of manoeuvring the forms clear of the structure to a point where they can be attached to the crane for final extraction, lifting and repositioning ready to receive the next concrete pour operation.
4. A means of providing a working platform at the external edge of the slab to eliminate the need for an independent scaffold which would be obstructive to the system.

The basic support members are usually a modified version of inverted adjustable steel props. These props, suitably braced and strutted, carry a framed decking which acts as the soffit formwork. To manoeuvre the forms into a position for attachment to the crane a framed wheeled arrangement can be fixed to the rear end of the tableform so that the whole unit can be moved forward with ease. The tableform is picked up by the crane at its centre of gravity by removing a loose centre board to expose the framework. The unit is then extracted clear of the structure, hoisted in the balanced horizontal position and lowered on to the recently cast slab for repositioning.

Another method, devised by Kwikform Ltd, uses a special lifting beam which is suspended from the crane at predetermined sling points which are lettered so that the correct balance for any particular assembly can be quickly identified, each table having been marked with the letter point required. The lifting beam is connected to the working platform attachment of the tableform so that when the unit formwork is lowered and then extracted by the crane the tableform and the lifting beam are in perfect balance. This method of system formwork is diagrammatically illustrated in Fig. II.62.

Students should appreciate that the selection of formwork for multistorey and complex structures is not a simple matter, but one which requires knowledge and experience in design appreciation, suitability of materials and site operations. Most of the large building contracting organisations have specialist staff for designing and detailing formwork. These contractors also have experienced and specialist site staff for the fabrication of formwork or if using a patent system to supervise the semi-skilled or unskilled labour being used. Badly designed and/or erected formwork can result in failure of the structure during the construction period or inaccurate and unacceptable members being cast, both of which can be financially disastrous for a company and ruinous to the firm's reputation.

14

Concrete surface finishes

The appearance of concrete members is governed by their surface finish and this is influenced by three major factors namely colour, texture and surface profile. The method used to produce the concrete member will have some degree of influence over the finish obtained since a greater control of quality is usually possible with precast concrete techniques under factory controlled conditions. Most precast concrete products can be cast in the horizontal position which again promotes better control over the resultant casting, whereas *in situ* casting of concrete walls and columns must be carried out using vertical casting techniques which not only has a lesser degree of control but also limits the types of surface treatments which can be successfully obtained direct from the mould or formwork.

COLOUR

The colour of concrete as produced direct from the mould or formwork depends upon the colour of the cement being used and to a lesser extent upon the colour of the fine aggregate. The usual methods of obtaining variations in the colour of finished concrete are:

1. *Using a coloured cement* — the range of colours available is limited and most are pastel shades; if a pigment is used to colour cement the cement content of the mix will need to be increased by approximately 10% to counteract the loss of strength due to the colouring additive.

2. *Using the colour of the coarse aggregate* — the outer matrix or cement paste is removed to expose the aggregate which not only imparts colour to the concrete surface but also texture.

Concrete can become stained during the construction period resulting in a mottled appearance. Some of the causes of this form of staining can be listed as follows:

1. *Using different quality timber within the same form* — generally timber which has a high absorption factor will give a darker concrete than timber with a low absorption factor. The same disfigurement can result from using old and new timber in the same form since older timber tends to give darker concrete than new timber with its probable higher moisture content and hence lower absorption.
2. *Formwork detaching itself from the concrete* — this allows dirt and dust to enter the space and attach itself to the green concrete surface.
3. *Type of release agent used* — generally the thinner the release agent used the better will be the result; even coating over the entire contact surface of the formwork or mould is also of great importance.

Staining occurring on mature concrete, usually after completion and occupation of the building, is very often due to bad design, poor selection of materials or poor workmanship. Large overhangs to parapets and sills without adequate throating can create damp areas which are vulnerable to algae or similar growths and pollution attack. Poor detailing of damp-proof courses can create unsightly stains by failing to fulfil their primary function of providing a barrier to dampness infiltration. Efflorescence can occur on concrete surfaces, although it is not so common as on brick walls. The major causes of efflorescence on concrete are allowing water to be trapped for long periods between the cast concrete and the formwork and poorly formed construction or similar joints allowing water to enter the concrete structure. Removal of efflorescence is not easy or always successful and therefore the emphasis should always be on good design and workmanship in the first instance. Methods of efflorescence removal range from wire brushing, various chemical applications to mechanical methods such as grit-blasting.

TEXTURE

When concrete first became acceptable as a substitute for natural stone in major building works the tendency was to try to recreate the smooth surface and uniform colour possible with natural stones. This kind of finish is difficult to achieve using the medium of concrete for the following reasons:

1. Natural shrinkage of concrete can cause hair line cracks on the surface.
2. Texture and colour can be affected by the colour of the cement, water content, degree of compaction and the quality of the formwork.
3. Pin holes on the surface can be caused by air being trapped between the concrete and the formwork.
4. Rough patches can result from the formwork adhering to the concrete face.
5. Grout leakage from the formwork can cause fins or honeycombing on the cast concrete.

Under site conditions it is not an easy task to keep sufficient control over the casting of concrete members to guarantee that these faults will not occur. A greater control is however possible with the factory type conditions prevailing in a well-organised precast concrete works. It should be noted that attempts to patch, mask or make good the above defects are nearly always visible.

Methods which can be used to improve the appearance of a concrete surface can be listed as follows:

1. Finishes obtained direct from the formwork or mould intended to conceal the natural defects by attracting the eye to a more obvious and visual point.
2. Special linings placed within the formwork to produce a smooth or profiled surface.
3. Removal of the surface matrix to expose the aggregate.
4. Applied surface finishes such as ceramic tiles, renderings and paints.

Formwork can be designed and constructed to highlight certain features such as the joints between form members or between concrete pours or lifts by adding to the inside of the form small fillets to form recessed joints, or conversely by recessing the formwork to form raised joints; the axiom being if you cannot hide or mask a joint make a feature of it. The use of sawn boards, to imprint the grain pattern, can give a visually pleasing effect to concrete surfaces particularly if boards of a narrow width are used.

A wide variety of textured, patterned and profiled surfaces can be obtained by using different linings within the formwork. Typical materials used are thermo-plastics, glass fibre mouldings, moulded rubber and PVC sheets; all of these materials can be obtained to form many various patterns giving many uses and are easily removed from the concrete surface. Glass fibre and thermo-plastic linings have the advantage of being capable of being moulded to any reasonable shape or profile. Ribbed and similar profiles can be produced by fixing materials such as troughed or corrugated steel into the form as a complete lining or as a partial lining to give a panelled effect. Coarse fabrics such as hessian can also be used to produce various textures providing the fabric is first soaked with a light mineral oil and squeezed to remove the excess oil before fixing as a precaution against the fabric becoming permanently embedded on the concrete surface.

The most common method of imparting colour and texture to a concrete surface is to expose the coarse aggregate which can be carried out by a variety of methods depending generally upon whether the concrete is green, mature, or yet to be cast. Brushing and washing is a common method employed on green concrete and is carried out using stiff wire or bristle brushes to loosen the surface matrix and clean water to remove the matrix and clean the exposed aggregate. The process should be carried out as soon as practicable after casting, which is usually within 2 to 6 hours of the initial pour but certainly not more than 18 hours after casting. When treating horizontal surfaces brushing should commence at the edges, whereas brushing to vertical surfaces should start at the base and be finished by washing from the top downwards. Skill is required with this method of exposing aggregates to ensure that only the correct amount of matrix is removed, the depth being dependent on the size of aggregate used. The use of retarding agents can be employed but it is essential that these are only used in strict accordance with the manufacturer's instructions.

Treatments which can be given to mature concrete to expose the aggregate include tooling, bush hammering and blasting techniques. These methods are generally employed when the depth of the matrix to be removed is greater than that usually associated with the brushing method described above. The most suitable age for treating the mature concrete surface depends upon such factors as the type of cement used and the conditions under which the concrete is cured, the usual period being from 2 to 8 weeks after casting.

The usual hand tooling methods which can be applied to natural stone such as point tooling and chiselling can be applied to a mature concrete surface but can be expensive if used for large areas. Tooling methods should not be used on concrete if gravel aggregates have been specified since these tend to shatter and leave unwanted pits on the surface. Hard point tooling near to sharp arrises should also be avoided because of the tendency to spalling at these edges; to overcome this problem a small plain margin at least 10 mm wide should be specified.

Except for small areas or where special treatment is required bush hammering has largely replaced the hand tooling techniques to expose the aggregate. Bush hammers are power-operated, hand-held tools to which two types of head can be used to give a series of light but rapid blows (e.g. 1 750 blows per minute) to remove about 3 mm of matrix for each passing. Circular heads with 21 cutting points are easier to control if working on confined or small areas than the more common and quicker roller head with its 90 cutting points.

Blasting techniques using sand, shot or grit are becoming increasingly popular since these methods do not cause spalling at the edges and generally result in a very uniform textured surface. Blasting can be carried out at almost any time after casting but preferably within 16 hours to 3 days of casting which gives a certain degree of flexibility to a programme of work. This method is of a specialist nature and is normally carried out by a sub-contractor who can usually expose between 4 to 6 m^2 of surface per hour. Sand is usually dispersed in a jet of water whereas grit is directed on to the concrete surface from a range of about 300 mm in a jet of compressed air and is the usual method employed. Shot is dispersed from a special tool with an enclosed head which collects and recirculates the shot.

Any method of removing the surface skin or matrix will reduce the amount of concrete cover over the reinforcement and therefore if this type of surface treatment is to be used an adequate cover of concrete should be specified such as 45 mm minimum before surface treatment.

When exposing aggregates by the above methods the finished result cannot be assured since the distribution of the aggregates throughout the concrete is determined by the way in which the fine and coarse aggregates disperse themselves during the mixing, placing and compacting operations. If a particular colour or type of exposed aggregate is required it will be necessary to use the chosen aggregate throughout the mix which may be both uneconomic and undesirable. A method which can be used for casting *in situ* concrete which will ensure the required distribution of aggregate over the surface and will enable a selected aggregate to be used on the surface is the aggregate transfer method.

This method entails sticking the aggregate to the rough side of pegboard sheets with a mixture of water-soluble cellulose compounds and sand fillers. The resultant mixture has a cream-like texture and is spread evenly over the pegboard surface to a depth of one third the aggregate size. The aggregate may be sprinkled over the surface and lightly tamped or alternatively placed by hand after which the prepared board is allowed to set and dry for about 36 hours. This prepared pegboard is used as a liner to the formwork with a loose baffle of hardboard or plywood immediately in front of the aggregate face as protection during the pouring of the cement. The baffle is removed as work proceeds. This is not an easy method of obtaining a specific exposed aggregate finish to *in situ* work and it is generally recommended that where possible simpler and just as effective precast methods should be considered.

Precast concrete casting can be carried out in horizontal or vertical moulds. Horizontal casting can have surface treatments carried out on the upper surface or the lower surface which is in contact with the mould base. Plain smooth surfaces can be produced by hand trowelling the upper surface but this is expensive in labour costs and requires a high degree of skill if a level surface of uniform texture and colour is to be obtained, whereas a good plain surface can usually be achieved direct from the mould using the lower face method. Various surface textures can be obtained from upper face casting by using patterned rollers, profiled tamping boards, using a straight edged tamping board drawn over the surface with a sawing action and by scoring the surface with brooms or rakes. Profiled finishes from lower face casting can be obtained by using a profiled mould base or a suitable base liner without any difficulty.

Exposed aggregate finishes can be obtained from either upper or lower face horizontal casting by spraying and brushing or by the tooling methods previously described. Alternatives for upper face casting are to trowel into the freshly cast surface selected aggregates or to trowel in a fine (10 mm) aggregate dry mortar mix and using a float with a felt pad to pick up the fine particles and cement leaving a clean exposed aggregate finish.

The sand bed method of covering the bottom of the mould with a layer of sand and placing the selected aggregate into the sand bed and immediately casting the concrete over the layer of aggregate is the usual method employed with lower face casting. Cast on finishes such as bricks, tiles and mosaic can be fixed in a similar manner to those described above for exposed aggregates. For all practical purposes vertical casting of precast concrete members needs the same considerations as those described previously in the context of *in situ* casting.

Another surface treatment which could be considered for concrete is grinding and polishing although this treatment is expensive and time consuming, being usually applied to surfaces such as slate, marble and terrazzo. The finish can be obtained by using a carborundum rotary grinder incorporating a piped water supply to the centre of the stone. The grinding operation should be carried out as soon as the mould or formwork has been removed by first wetting the surface and then grinding, allowing the stone dust and water to mix forming an abrasive paste. The operation is completed by washing and brushing to remove the paste residue. Dry grinding is possible but this creates a great deal of dust from which the operative's eyes and lungs must be protected by wearing protective masks.

It is very often wrongly assumed that by adding texture or shape to a concrete surface any deficiency in design, formwork fabrication or workmanship can be masked. This is not so; a textured surface, like the plain surface, will only be as good as the design and workmanship involved in producing the finished component and any defects will be evident with all finishes.

15
Precast concrete frames

The overall concept of a precast concrete frame is the same as any other framing material. Single or multi-storey frames can be produced on the skeleton or box frame principle. Single and two-storey buildings can also be produced as portal frames, a method generally reserved for advanced level study. Most precast concrete frames are produced as part of a 'system' building and therefore it is only possible to generalise in an overall study of this method of framing.

Advantages

1. Mixing, placing and curing of the concrete carried out under factory-controlled conditions which results in uniform and accurate units. The casting, being an 'off site' activity, will release site space which would have been needed for the storage of cement and aggregates, mixing position, timber store and fabrication area for formwork and the storage, bending and fabrication of the reinforcement.
2. Repetitive standard units reduce costs: it must be appreciated that the moulds used in precast concrete factories are precision made, resulting in high capital costs. These costs must be apportioned over the number of units to be cast.
3. Frames can be assembled on site in cold weather which helps with the planning, programming and progressing of the building operations. This is important to the contractor since delays can result in the monetary penalty clauses, for late completion of the contract, being invoked.

4. In general the frames can be assembled by semi-skilled labour. With the high turnover rate of labour within the building industry operatives can be recruited and quickly trained to carry out these activities.

Disadvantages

1. System building is less flexible in its design concept than purpose-made structures. It must be noted that there is a wide variety of choice of systems available to the designer, so that most design briefs can be fulfilled without too much modification to the original concept.
2. Mechanical lifting plant will be needed to position the units; this can add to the overall contracting costs since generally larger plant is required for precast concrete structures than for *in situ* concrete structures.
3. Programming may be restricted by controls on delivery and unloading times laid down by the police. Restrictions on deliveries is a point which must be established at the tender period so that the tender programme can be formulated with a degree of accuracy and any overtime payments can be included in the unit rates for pricing.
4. Structural connections between the precast concrete units can present both design and contractual problems. The major points to be considered are protection against weather, fire and corrosion, appearance and the method of construction. The latter should be issued as an instruction to site, setting out in detail the sequence, temporary supports required and full details of the joint.

METHODS OF CONNECTIONS

Foundation connections

Precast columns are connected to their foundations by one of two methods, depending mainly upon the magnitude of the load. For light and medium loads the foot of the column can be placed in a pocket left in the foundation. The column can be plumbed and positioned by fixing a collar around its perimeter and temporarily supporting the column from this collar by using raking adjustable props. Wedges can be used to give added rigidity whilst the column is being grouted into the pocket (see Fig. II.63). The alternative method is to cast or weld on a base plate to the foot of the column and use holding down bolts to secure the column to its foundation in the same manner as described in detail for structural steelwork (see Fig. II.63).

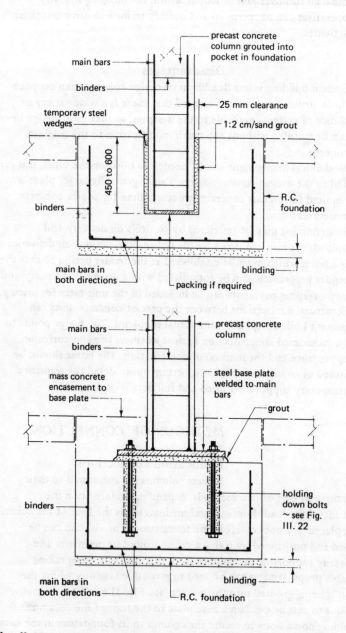

Fig. II.63 P.C.C. column to foundation connections

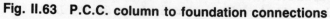

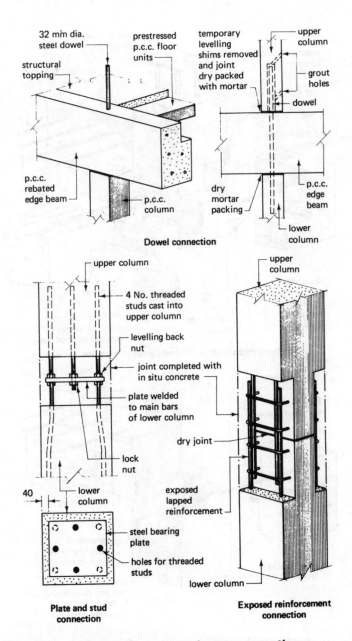

Dowel connection

Plate and stud connection

Exposed reinforcement connection

Fig. II.64 Precast concrete column connections

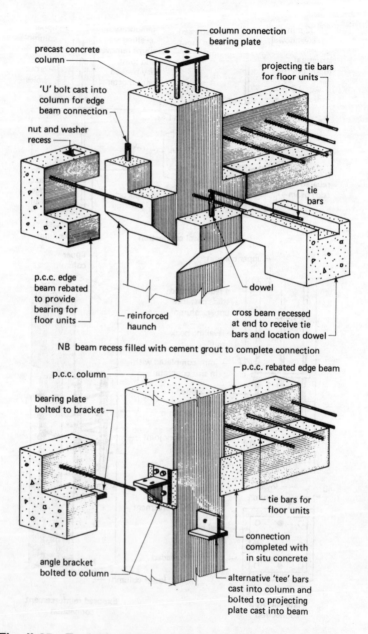

column connection
bearing plate

precast concrete
column

projecting tie bars
for floor units

'U' bolt cast into
column for edge
beam connection

nut and washer
recess

tie
bars

p.c.c. edge
beam rebated
to provide
bearing for
floor units

reinforced
haunch

dowel

cross beam recessed
at end to receive tie
bars and location dowel

NB beam recess filled with cement grout to complete connection

p.c.c. column

p.c.c. rebated edge beam

bearing plate
bolted to bracket

tie bars for
floor units

connection
completed with
in situ concrete

angle bracket
bolted to column

alternative 'tee' bars
cast into column and
bolted to projecting
plate cast into beam

Fig. II.65 Typical precast concrete beam connections

Column connections

The main principle involved in making column connections is to ensure continuity and this can be achieved by a variety of methods. In simple connections a direct bearing and grouted dowel joint can be used, the dowel being positioned in the upper or lower column. Where continuity of reinforcement is required the reinforcement from both upper and lower columns left exposed and either lapped or welded together before completing the connection with *in situ* concrete. A more complex method is to use a stud and plate connection where one set of threaded bars are connected through a steel plate welded to a set of bars projecting from the lower column; again the connection is completed with *in situ* concrete. Typical column connections are shown in Fig. II.64. Column connections should be made at floor levels but above the beam connections, a common dimension being 600 mm above structural floor level. The columns can be of single or multi-storey height, the latter having provisions for beam connections at the intermediate floor levels.

Beam connections

As with columns, the main emphasis is on continuity within the joint. Three basic methods are used:

1. A projecting concrete haunch is cast on to the column with a locating dowel or stud bolt to fix the beam.
2. A projecting metal corbel is fixed to the column and the beam is bolted to the corbel.
3. Column and beam reinforcement, generally in the form of hooks, are left exposed. The two members are hooked together and covered with *in situ* concrete to complete the joint.

With most beam to column connections lateral restraint is provided by leaving projecting reinforcement from the beam sides to bond with the floor slab or precast concrete floor units (see Fig. II.65).

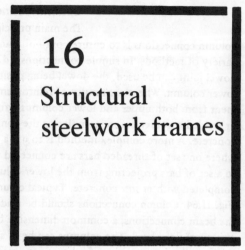

16
Structural
steelwork frames

Structural steel as a means of constructing a
framed building has been used since the beginning of the twentieth century
and was the major structural material used until the advent of the Second
World War, which led to a shortage of the raw material. This shortage led
to an increase in the use of *in situ* and precast concrete frames. Today
both systems are used and this means a comparison must be made before
any particular framing medium is chosen. The main factors to be
considered in making this choice are:

Site costs: a building owner will want to obtain a financial return on his
capital investment as soon as possible, therefore speed of construction is of
paramount importance. The use of a steel or precast concrete frame will
enable the maximum amount of prefabrication off site, during which time
the general contractor can be constructing the foundations in preparation
for the erection of the frame. To obtain the maximum utilisation of a site
the structure needs to be designed so that the maximum amount of
rentable floor area is achieved. Generally prefabricated section sizes are
smaller than comparable·*in situ* concrete members, due mainly to the
greater control over manufacture obtainable under factory conditions and
thus these will occupy less floor area.

Construction costs: the main factors are design considerations, availability
of labour, availability of materials and site conditions. Concrete is a
flexible material which allows the designer to be more creative than
working within the rigid confines of standard steel sections. However, as
the complexity of shape and size increases so does the cost of formwork

160

and for the erection of a steel structure skilled labour is required, whereas activities involved with precast concrete structures can be carried out by the more readily available semi-skilled labour working under the direction of a competent person. The availability of materials fluctuates and only a study of current market trends can give an accurate answer to this problem. Site conditions regarding storage space, fabrication areas and manoeuvrability around and over the site can well influence the framing method chosen.

Maintenance costs: these can be considered in the short or long term but it is fair to say that in most framed buildings the costs are generally negligible if the design and workmanship is sound. Steelwork, because of its corrosive properties, will need some form of protective treatment but since most steel structures have to be given a degree of fire resistance the fire protective method may well perform the dual function.

STRUCTURAL STEEL FRAMES
The design, fabrication, supply and erection of a structural steel frame is normally placed in the hands of a specialist sub-contractor. The main contractor's task is to provide the foundation bases in the correct positions and to the correct levels with the necessary holding down fixing bolts. The designer will calculate the loadings, stresses and reactions in the same manner as for reinforced concrete and then select a standard steel member whose section properties meet the design requirements.

Standard steel sections are given in BS 4 and BS 4848 and in the *Handbook on Structural Steelwork* published jointly by the British Constructional Steelwork Association Limited and the Constructional Steel Research and Development Organisation, which gives the following section types:

Universal beams: these are a range of sections supplied with tapered or parallel flanges and are designated by their serial size x mass in kilograms per metre run. To facilitate the rolling operation of universal beam sections the inner profile is a constant dimension for any given serial size. The serial size is therefore only an approximate width and breadth and is given in millimetres.

Joists: a range of small size beams which have tapered flanges and are useful for lintels and small frames around openings. In the case of joists the serial size is the overall nominal dimension.

Universal columns: these members are rolled with parallel flanges and are designated in the same manner as universal beams. It is possible to design a column section to act as a beam and conversely a beam section to act as a column.

Channels: rolled with tapered flanges and designated by their nominal overall dimension x mass per metre run and can be used for trimming and bracing members or as a substitute for joist sections.

Angles: light framing and bracing sections with parallel flanges. The flange or leg lengths can be equal or unequal and the sections are designated by the nominal overall leg lengths x nominal thickness of the flange.

T bars: used for the same purposes as angles and are available as rolled sections with a short or long stalk or alternatively they can be cut from a standard universal beam or column section. Designation is given by the nominal overall breadth and depth x mass per metre run.

Typical standard steel sections are shown in Fig. II.66.

CASTELLATED UNIVERSAL SECTIONS

These are formed by flame cutting a standard universal beam or column section along a castellated line; the two halves so produced are welded together to form an open web beam. The resultant section is one and a half times the depth of the section from which it was cut (see Fig. II.67). This increase in depth gives greater resistance to deflection without adding extra weight but will reduce the clear head-room under the beams unless the overall height of the building is increased. Castellated sections are economical when used to support lightly loaded floor or roof slabs and the voids in the web can be used for housing services. With this form of beam the shear stresses at the supports can be greater than the resistance provided by the web; in these cases one or two voids are filled in by welding into the voids metal blanks.

Connections

Connections in structural steelwork are classified as either shop connections or site connections and can be made by using bolts, rivets or by welding.

Bolts

Black bolts: the cheapest form of bolt available, the black bolt can be either hot or cold forged, the thread being machined onto the shank. The allowable shear stresses for this type of bolt are low and therefore they should only be used for end connections of secondary beams or in

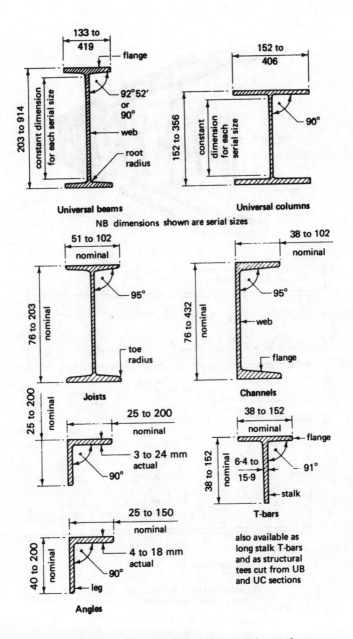

Fig. II.66 Typical BS 4 and BS 4848 steel sections

163

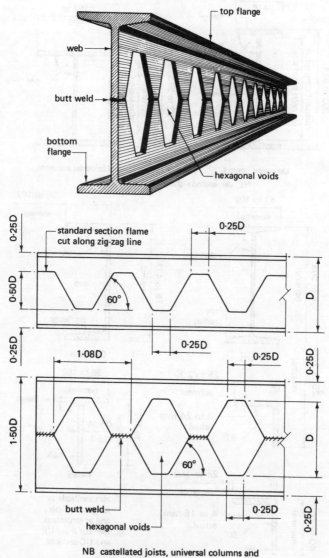

NB castellated joists, universal columns and
zed sections also available

Fig. II.67 Castellated beams

164

conjunction with a seating cleat which has been designed and fixed to resist all the shear forces involved. The clearance in the hole for this form of bolt is usually specified as 1.6 mm over the diameter of the bolt. The term black bolts does not necessarily indicate the colour but is the term used to indicate the comparatively wide tolerances to which these products are usually made. BS 4190 gives recommendations for black bolts and nuts for a diameter range of 5 to 68 mm inclusive.

Bright bolts: these have a machined shank and are therefore of greater dimensional accuracy fitting into a hole with a small clearance allowance. The stresses allowed are similar to those permitted for rivets. Bright bolts are sometimes called turned and fitted bolts.

High strength friction bolts: manufactured from high tensile steels and are used in conjunction with high tensile steel nuts and tempered washers. These bolts have generally replaced rivets and black bolts for both shop and site connections since fewer bolts are needed and hence the connection size is reduced. The object of this form of bolt is to tighten it to a predetermined shank tension in order that the clamping force thus provided will transfer the loads in the connecting members by friction between the parts and not by shear in or bearing on the bolts. Generally a torque controlled spanner or pneumatic impact wrench is used for tightening; other variations to ensure the correct torque are visual indicators such as a series of pips under the head or washer which are flattened when the correct amount of shank tension has been reached. Nominal standard diameters available are from 12 to 36 mm with lengths ranging from 40 to 500 mm, as recommended in BS 4395.

The holes to receive bolts should always be drilled in a position known as the back mark of the section. The back mark is the position on the flange where the removal of material to form a bolt or rivet hole will have the least effect upon the section properties. Actual dimensions and recommended bolt diameters are given in the *Handbook on Structural Steelwork*.

Rivets
Made from mild steel to the recommendations of BS 4620 rivets have been generally superseded by bolted and welded connections for structural steel frames. Rivets are available as either cold or hot forged with a variety of head shapes ranging from an almost semi-circular or snap head to a countersunk head for use when the projection of a snap, universal or flat head would create an obstruction. Small diameter rivets can be cold driven but the usual practice is to drive rivets whilst they are hot. Rivets,

like bolts, should be positioned on the back mark of the section; typical spacings are 2½ diameters centre to centre and 1¾ diameters from the end or edge to the centre line of the first rivet.

Welding

Primarily considered as a shop connection since the cost together with the need for inspection, which can be difficult on site, generally makes this method uneconomic for site connections.

The basic methods of welding are oxy-acetylene and electric arc. A blowpipe is used for oxy-acetylene which allows the heat from the burning gas mixture to raise the temperature of the surfaces to be joined. A metal filler rod is held in the flame and the molten metal from the filler rod fuses the surfaces together.

In the alternative method an electric arc is struck between a metal rod connected to a suitable low voltage electrical supply and the surface to be joined which must be earthed or resting on an earthed surface. The heat of the arc causes the electrode or metal rod to melt and the molten metal can be deposited in layers to fuse the pieces to be joined together. With electrical arc welding the temperature rise is confined to the local area being welded whereas oxy-acetylene causes a rise in metal temperature over a general area.

Welds are classified as either fillet or butt welds. Fillet welds are used to the edges and ends of members and forms a triangular fillet of welding material. Butt welds are used on chamfered end to end connections.

Structural steel connections

Base connections: are of one or two forms, the slab or bloom base and the gusset base. In both methods a steel base plate is required to spread the load of the column on to the foundation. The end of the column and the upper surface of the base plate should be machined to give a good inter-face contact when using a bloom base. The base plate and column can be connected together by using cleats or by fillet welding (see Fig. II.68).

The gusset base is composed of a number of members which reduce the thickness of the base plate and can be used to transmit a high bending moment to the foundations. A machined interface between column and base plate will enable all the components to work in conjunction with one another, but if this method is not adopted the connections must transmit all the load to the base plate (see Fig. II.68). The base is joined to the foundation by holding down bolts which must be designed to resist the uplift and tendency of the column to overturn. The bolt diameter, bolt length and size of plate washer are therefore important. To allow for fixing tolerances the bolts are initially housed in a void or pocket which is filled with grout at the same time as the base is grouted on to the foundation. To level and plumb the columns steel wedges are inserted between the underside of the base plate and the top of the foundation (see Fig. II.68).

Beam to column connections: these can be designed as simple connections where the whole of the load is transmitted to the column through a seating cleat. This is an expensive method requiring heavy sections to overcome deflection problems. The usual method employed is the semi-rigid connection where the load is transmitted from the beam to the column by means of top cleats and/or web cleats; for ease of assembly an erection cleat on the underside is also included in the connection detail (see Fig. II.69). A fully rigid connection detail, which gives the greatest economy on section sizes, is made by welding the beam to the column (see Fig. II.69). The uppermost beam connection to the column can be made by the methods described above or alternatively a bearing connection can be used, which consists of a cap plate fixed to the top of the column to which the beams can be fixed either continuously over the cap plate or with a butt joint (see Fig. II.69).

Column splices: these are made at floor levels but above the beam connections. The method used will depend upon the relative column sections (see Fig. II.70).

Beam to beam connections: the method used will depend upon the relative depths of the beams concerned. Deep beams receiving small secondary beams can have a shelf angle connection whereas other depths will need to be connected by web cleats (see Fig. II.71).

FRAME ERECTION

This operation will not normally be commenced until all the bases have been cast and checked since the structural steelwork contractor will need a clear site for manoeuvring the steel members into position. The usual procedure is to erect two storeys of steelwork before final plumbing and levelling takes place.

The grouting of the base plates and holding down bolts is usually left until the whole structure has been finally levelled and plumbed. The grout is a neat cement or cement/sand mixture depending on the gap to be filled:

12 to 25 mm gap — stiff mix of neat cement;
25 to 50 mm gap — fluid mix of 1 : 2 cement/sand and tamped;
Over 50 mm gap — stiff mix of 1 : 2 cement/sand and rammed.

With large base plates a grouting hole is sometimes included but with smaller plates three sides of the base plate are sealed with puddle clay,

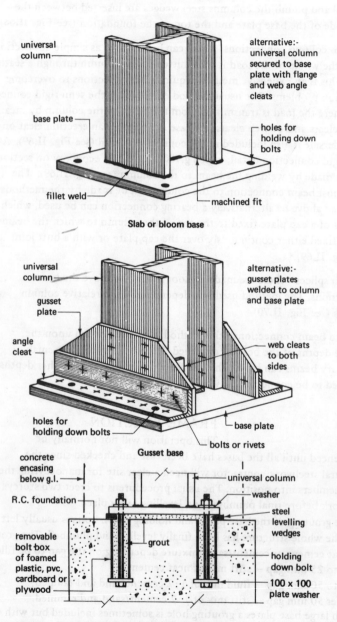

Fig. II.68 Structural steel column bases

Labels in figure:

- universal column
- alternative:- universal column secured to base plate with flange and web angle cleats
- base plate
- holes for holding down bolts
- fillet weld
- machined fit

Slab or bloom base

- universal column
- alternative:- gusset plates welded to column and base plate
- gusset plate
- angle cleat
- web cleats to both sides
- holes for holding down bolts
- bolts or rivets
- base plate

Gusset base

- concrete encasing below g.l.
- R.C. foundation
- removable bolt box of foamed plastic, pvc, cardboard or plywood
- grout
- universal column
- washer
- steel levelling wedges
- holding down bolt
- 100 x 100 plate washer

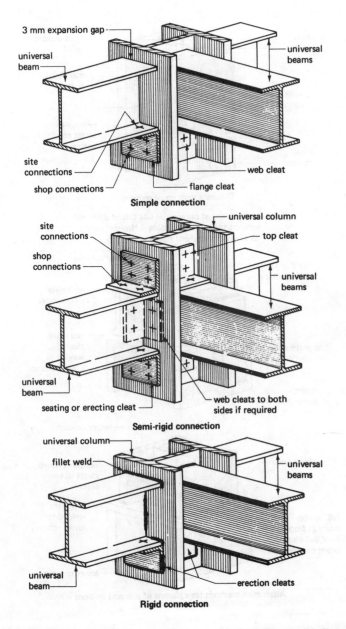

Fig. II.69 Structural steel beam to column connections

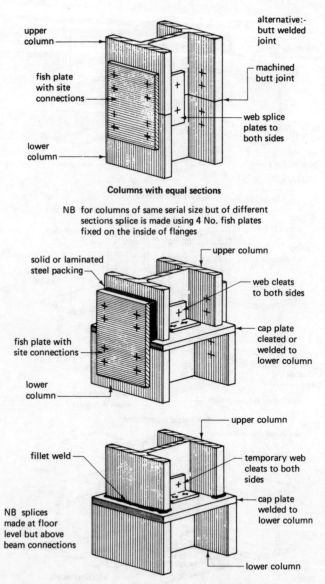

upper column

alternative:-
butt welded
joint

fish plate
with site
connections

machined
butt joint

web splice
plates to
both sides

lower
column

Columns with equal sections

NB for columns of same serial size but of different
sections splice is made using 4 No. fish plates
fixed on the inside of flanges

solid or laminated
steel packing

upper column

web cleats
to both sides

fish plate with
site connections

cap plate
cleated or
welded to
lower column

lower
column

upper column

fillet weld

temporary web
cleats to both
sides

NB splices
made at floor
level but above
beam connections

cap plate
welded to
lower column

lower column

Alternative methods for columns of unequal sections

Fig. II.70 Structural steel column splices

170

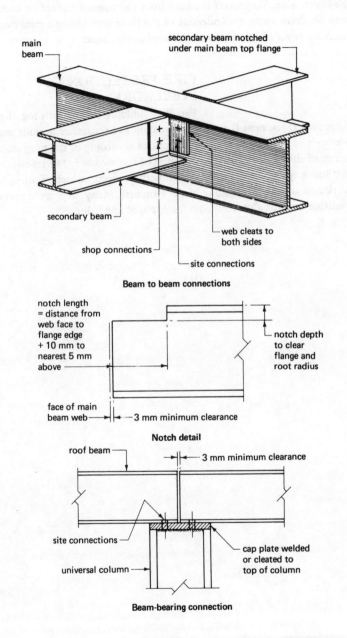

main beam

secondary beam notched under main beam top flange

web cleats to both sides

shop connections

site connections

secondary beam

Beam to beam connections

notch length
= distance from
web face to
flange edge
+ 10 mm to
nearest 5 mm
above

notch depth
to clear
flange and
root radius

face of main
beam web

3 mm minimum clearance

Notch detail

roof beam

3 mm minimum clearance

site connections

cap plate welded
or cleated to
top of column

universal column

Beam-bearing connection

Fig. II.71 Structural steel beam to beam connections

bricks or formwork and the grout introduced through the open edge on the fourth side. To protect the base from corrosion it should be encased with concrete up to the underside of the floor level giving a minimum concrete cover of 75 mm to all the steel components.

FIRE PROTECTION OF STEELWORK

Part B of Building Regulations together with Approved Document B gives the minimum fire resistance periods and methods of protection for steel structures according to the purpose group of the building and the function of the member. The traditional method is to encase the steel section with concrete, which requires form-work and adds to the loading of the structure. Many 'dry' techniques are available but not all are suitable for exposed conditions.

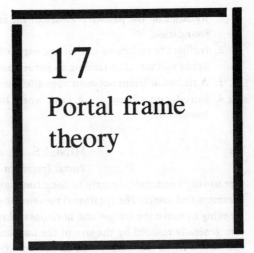

17
Portal frame theory

             A portal frame may be defined as a
continuous or rigid frame which has the basic characteristic of a rigid or
restrained joint between the supporting member or column and the
spanning member or beam. The object of this continuity of the portal
frame is to reduce the bending moment in the spanning member by
allowing the frame to act as one structural entity, thus distributing the
stresses throughout the frame.

  If a conventional simply supported beam was used (over a large span)
an excessive bending moment would occur at mid-span which would
necessitate a deep heavy beam or a beam shaped to give a large cross
section at mid-span. Alternatively a deep cross member of lattice struts
and ties could be used. The main advantage of the simply supported frame
lies in the fact that the column loading is for all practicable purposes axial
and therefore no bending is induced into the supporting members, which
may well ease design problems since it would be statically determinate, but
does not necessarily produce an economic structure. Furthermore the use
of a portal frame eliminates the need for a lattice of struts and ties within
the roof space, giving a greater usable volume to the structure and
generally a more pleasing internal appearance.

  The transfer of stresses from the beam to the column in rigid frames
will require special care in the design of the joint between the members,
similarly the horizontal thrust and/or rotational movement at the
foundation connection needs careful consideration. Methods used to
overcome excessive forces at the foundation are:

1. Reliance on the passive pressure of the soil surrounding the foundation.
2. Inclined foundations so that the curve of pressure is normal to the upper surface, thus tending to induce only compressive forces.
3. A tie bar or beam between opposite foundations.
4. Introducing a hinge or pin joint where the column connects to the foundation.

HINGES

Portal frames of moderate height and span are usually connected directly to their foundation bases forming rigid or unrestrained joints. The rotational movement caused by wind pressures tending to move the frames and horizontal thrusts of the frame loadings are generally resisted by the size of the base and the passive earth pressures. When the frames start to exceed 4.000 m in height and 15.000 m in span the introduction of a hinged or pin joint at the base connection should be considered.

A hinge is a device which will allow free rotation to take place at the point of fixity but at the same time will transmit both load and shear from one member to another. They are sometimes called pin joints, unrestrained joints and non-rigid joints. Since no bending moment is transmitted through a hinged joint the design is simplified by the structural connection becoming statically determinate. In practice it is not always necessary to provide a true 'pivot' where a hinge is included but to provide just enough movement to ensure the rigidity at the connection is low enough to overcome the tendency of rotational movement.

Hinges can be introduced into a portal frame design at the base connections and at the centre or apex of the spanning member, giving three basic forms of portal frame:

1. *Fixed or rigid portal frame* — all connections between frame members are rigid. This will give bending moments of lower magnitude more evenly distributed than other forms. Used for small to medium size frames where the moments transferred to the foundations will not be excessive.
2. *Two pin portal frame* — hinges are used at the base connections to eliminate the tendency of the base to rotate. The bending moments resisted by the supporting members will be greater than those encountered in the rigid portal frame. Main use is where high base moments and weak ground conditions are encountered.
3. *Three pin portal frame* — this form of frame has hinged joints at the base connections and at the centre of the spanning member. The

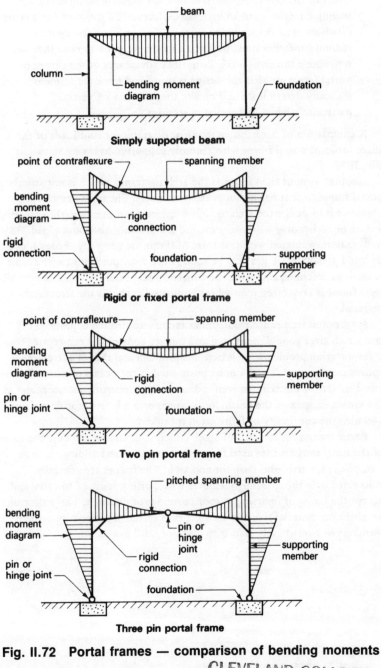

Simply supported beam

beam
column
bending moment diagram
foundation

Rigid or fixed portal frame

point of contraflexure
spanning member
bending moment diagram
rigid connection
rigid connection
foundation
supporting member

Two pin portal frame

point of contraflexure
spanning member
bending moment diagram
rigid connection
pin or hinge joint
foundation
supporting member

Three pin portal frame

pitched spanning member
bending moment diagram
pin or hinge joint
rigid connection
pin or hinge joint
foundation
supporting member

Fig. II.72 Portal frames — comparison of bending moments

effect of the third hinge is to reduce the bending moments in the spanning member but to increase deflection. To overcome this latter disadvantage a deeper beam must be used or alternatively the spanning member must be given a moderate pitch to raise the apex well above the eaves level. Two other advantages of the three pin portal frame are that the design is simplified since the frame is statically determinate and on site they are easier to erect, particularly when preformed in sections.

A comparison of the bending moment diagrams for roof loads of the three forms of portal frame with a simply supported beam are shown in Fig. II.72.

Another form of rigid frame is the arch rib frame, which is not strictly a portal frame since it has no supporting members. The main design objective is to design the arch to follow the curve of pressure, thus creating a state of no bending when subjected to a uniformly distributed load. Any moments encountered with this form of frame are generally those induced by wind pressures. Hinges may be used in the same positions for the same reasons as described above for the conventional portal frames. The arch rib rigid frame is very often used where laminated timber is the structural material.

Most portal frames are made under factory controlled conditions off site which gives good dimensional and quality control but can create transportation problems. To lessen this problem and that of site erection splices may be used. These can be positioned at the points of contraflexure (see Fig. II.72), junction between spanning and supporting members and at the crown or apex of the beam. Most hinges or pin joints provide a point at which the continuity of fabrication is broken.

Portal frames constructed of steel, concrete or timber can take the form of the usual roof profiles used for single or multi-span buildings such as flat, pitched, northlight, monitor and arch. The frames are generally connected over the spanning members with purlins designed to carry and accept the fixing of lightweight roof coverings or deckings. The walls can be of similar material fixed to sheeting rails attached to the supporting members or alternatively clad with brick or infill panels.

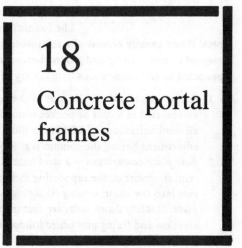

18

Concrete portal frames

Concrete portal frames are invariably manufactured from high quality precast concrete suitably reinforced. In common with all precast concrete components for buildings, rapid advances in design and use were made after the Second World War due mainly to the shortage of steel and timber which prevailed at that time. In the main the use of precast concrete portal frames is confined to low pitch (4° to 22½°) single span frames but two storey and multi-span frames are available, giving a wide range of designs from only a few standard components.

The frames are generally designed to carry a lightweight (34 kg/m² maximum) roof sheeting or decking fixed to precast concrete purlins. Most designs have an allowance for snow loading of up to 73 kg/m² in addition to that allowed for the dead load of the roof covering. Wall finishes can be varied and intermixed since they are non-load bearing and therefore have to provide only the degree of resistance required for fire, thermal and sound insulation, act as a barrier to the elements and resist positive and negative wind pressures. Sheet claddings are fixed in the traditional manner, using hook bolts and purlins; sheet wall claddings are fixed in a similar manner to sheeting rails of precast concrete or steel spanning between or over the supporting members. Brick or block wall panels either of solid or cavity construction can be built off a ground beam constructed between the foundation bases or alternatively they can be built off the ground floor slab. It must be remembered that all such claddings must comply with any relevant Building Regulations.

FOUNDATIONS AND FIXINGS

The foundations for a precast concrete portal frame usually consist of a reinforced concrete isolated base or pad designed to suit loading and ground bearing conditions. The frame can be connected to the foundations by a variety of methods:

1. *Pocket connection* — the foot of the supporting member is located and housed in a void or pocket formed in the base so that there is an all round clearance of 25 mm to allow for plumbing and final adjustment before the column is grouted into the foundation base.
2. *Base plate connection* — a steel base plate is welded to the main reinforcement of the supporting member, or alternatively it could be cast into the column using fixing lugs welded to the back of the base plate. Holding down bolts are cast into the foundation base; the erection and fixing procedure follows that described for structural steelwork (see Chapter 16).
3. *Pin joint or hinge connection* — a special base or bearing plate is bolted to the foundation and the mechanical connection is made when the frames are erected — see Fig. II.76.

The choice of connection method depends largely upon the degree of fixity required and the method adopted by the manufacturer for his particular system.

ADVANTAGES

The main advantages of using precast concrete portal frames can be enumerated thus:

1. Factory production will result in accurate and predictable components since the criteria for design, quality and workmanship recommended in BS 8110 can be more accurately controlled under factory conditions than casting components *in situ.*
2. Most manufacturers produce a standard range of interchangeable components which, within the limitations of their systems, gives a well-balanced and flexible design range covering most roof profiles, single span frames, multi-span frames and lean-to roof attachments. By adopting this limited range of members the producers of precast portal frames can offer their products at competitive rates coupled with reasonable delivery periods.
3. Maintenance of precast concrete frames is not usually required unless the building owner chooses to paint or clad the frames.
4. Precast concrete products have their own built-in natural resistance to fire and therefore no fire-resistant treatment is required. By

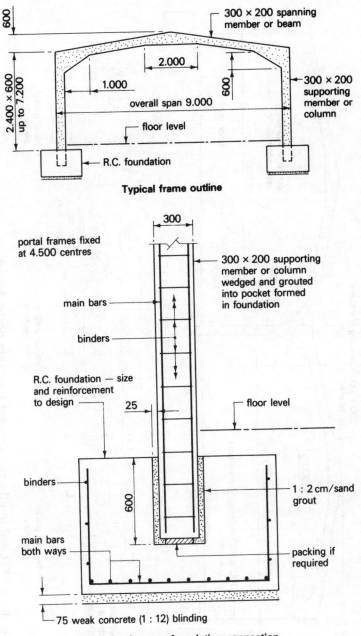

600

300 × 200 spanning
member or beam

2.000

1.000

600

2.400 × 600
up to 7.200

overall span 9.000

300 × 200
supporting
member or
column

floor level

R.C. foundation

Typical frame outline

300

portal frames fixed
at 4.500 centres

300 × 200 supporting
member or column
wedged and grouted
into pocket formed
in foundation

main bars

binders

R.C. foundation — size
and reinforcement
to design

25

floor level

binders

600

1 : 2 cm/sand
grout

main bars
both ways

packing if
required

75 weak concrete (1 : 12) blinding

Typical column to foundation connection

Fig. II.73 Typical single span pcc frame

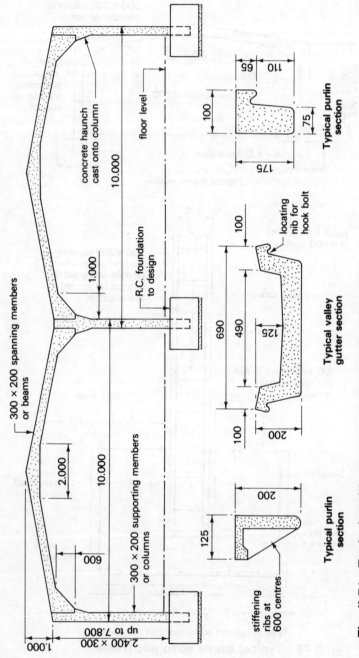

Fig. II.74 Typical multi-span precast concrete portal frame

concrete haunch cast onto column

floor level

R.C. foundation to design

Typical purlin section

Typical valley gutter section

locating nib for hook bolt

300 × 200 spanning members or beams

2.000

300 × 200 supporting members or columns

stiffening ribs at 600 centres

Typical purlin section

up to 7.800

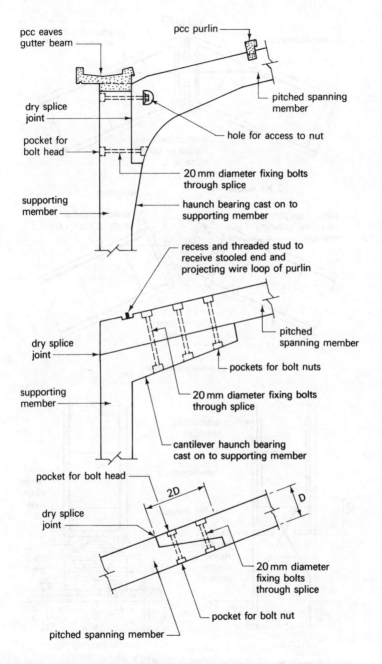

pcc eaves gutter beam

pcc purlin

dry splice joint

pocket for bolt head

supporting member

pitched spanning member

hole for access to nut

20 mm diameter fixing bolts through splice

haunch bearing cast on to supporting member

recess and threaded stud to receive stooled end and projecting wire loop of purlin

dry splice joint

supporting member

pitched spanning member

pockets for bolt nuts

20 mm diameter fixing bolts through splice

cantilever haunch bearing cast on to supporting member

pocket for bolt head

dry splice joint

2D

D

20 mm diameter fixing bolts through splice

pocket for bolt nut

pitched spanning member

Fig. II.75 Typical splice details for pcc portal frames

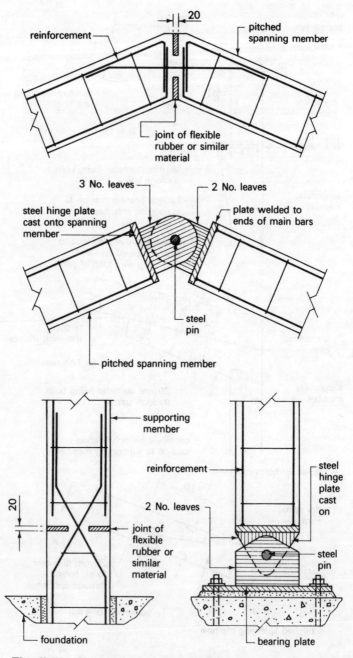

Fig. II.76 Typical hinge details for pcc portal frames

varying the cover of concrete over the reinforcement most frames up to 24.000 m span are given a 1-hour fire resistance and frames exceeding this span are rated at 2-hour fire resistance.

5. The wind resistance of precast concrete portal frames to both positive and negative pressures is such that wind bracing is not usually required.
6. Where members of the frame are joined or spliced together the connections are generally mechanical (nut and bolt) and therefore the erection and jointing can be carried out by quickly trained semi-skilled labour.
7. The clean lines of precast concrete portal frames are considered to be aesthetically pleasing.
8. In most cases the foundation design, setting out and construction can be carried out by the portal frame sub-contracting firm.

Typical details of single span frames, multi-span frames, cladding supports, splicing and hinges are shown in Figs. II.73 to II.76.

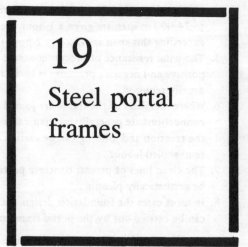

19
Steel portal frames

Steel portal frames can be fabricated from standard universal beam, column and box sections. Alternatively a lattice construction of flats, angles or tubulars can be used. Most forms of roof profiles can be designed and constructed giving a competitive range when compared with other materials used in portal frame construction. The majority of systems employ welding techniques for the fabrication of components which are joined together on site using bolts or welding. An alternative system uses special knee joint, apex joint and base joint components which are joined on site to square cut standard beam or column sections supplied by the main contractor or by the manufacturer producing the jointing pieces.

The frames are designed to carry lightweight roof coverings of the same loading conditions as those given previously for precast concrete portal frames. Similarly wall claddings can be of the same specification as for precast concrete portal frames and fixed in the same manner. Any relevant Building Regulations must be observed and if the usage of the building, irrespective of the framing material, is for an industrial process the roof would have to comply with the requirements of The Building Regulations 1985 Schedule 1, Part L (see Part V — roofs).

FOUNDATIONS AND FIXINGS

The foundation is usually a reinforced concrete isolated base or pad foundation designed to suit loading and ground bearing conditions. The connection of the frame to the foundation can be by one of three basic methods:

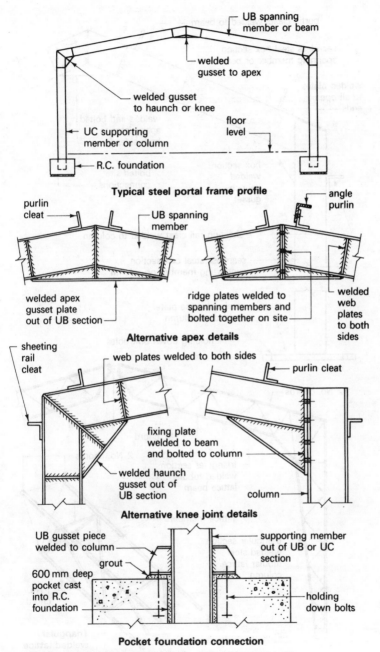

Typical steel portal frame profile

purlin cleat

UB spanning member

angle purlin

welded apex gusset plate out of UB section

ridge plates welded to spanning members and bolted together on site

welded web plates to both sides

Alternative apex details

sheeting rail cleat

web plates welded to both sides

purlin cleat

fixing plate welded to beam and bolted to column

welded haunch gusset out of UB section

column

Alternative knee joint details

UB gusset piece welded to column

supporting member out of UB or UC section

grout

600 mm deep pocket cast into R.C. foundation

holding down bolts

Pocket foundation connection

Fig. II.77 Typical steel portal frame details

185

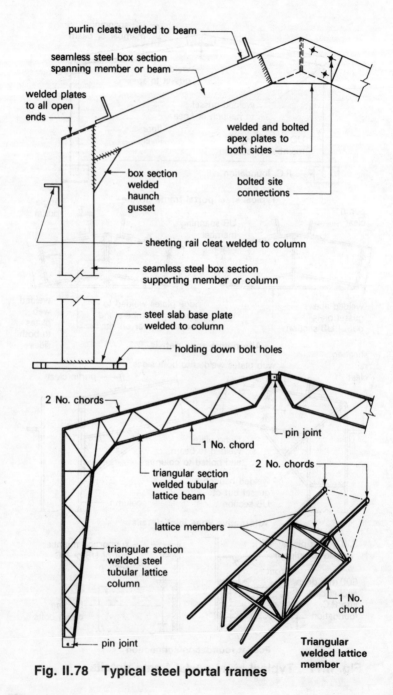

purlin cleats welded to beam

seamless steel box section spanning member or beam

welded plates to all open ends

welded and bolted apex plates to both sides

box section welded haunch gusset

bolted site connections

sheeting rail cleat welded to column

seamless steel box section supporting member or column

steel slab base plate welded to column

holding down bolt holes

2 No. chords

pin joint

1 No. chord

triangular section welded tubular lattice beam

2 No. chords

lattice members

triangular section welded steel tubular lattice column

1 No. chord

pin joint

Triangular welded lattice member

Fig. II.78 Typical steel portal frames

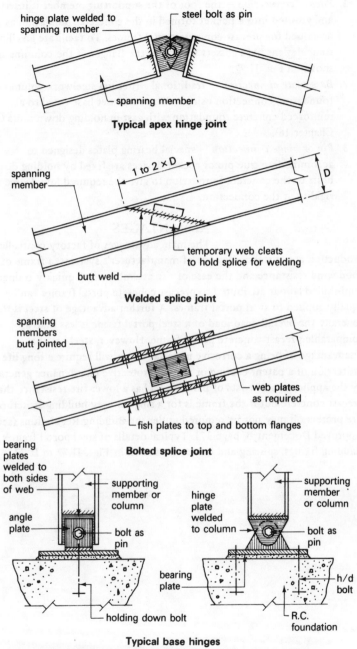

Typical apex hinge joint

Welded splice joint

Bolted splice joint

Typical base hinges

Fig. II.79   Steel portal frames — splices and hinges

187

1. *Pocket connection* — the foot of the supporting member is inserted and grouted into a pocket formed in the concrete foundation as described for precast concrete portal frames. To facilitate levelling some designs have gussets welded to the flanges of the columns as shown in Fig. II.77.
2. *Base plate connection* — traditional structural steelwork column to foundation connection using a slab or a gusset base fixed to a reinforced concrete foundation with cast in holding down bolts (see Chapter 16).
3. *Pin or hinge connection* — special bearing plates designed to accommodate true pin or rocker devices are fixed by holding down bolts to the concrete foundation to give the required low degree of rigidity at the connection.

ADVANTAGES

The main advantages of factory controlled production are: a standard range of manufacturer's systems, a frame of good wind resistance and the ease of site assembly using quickly trained semi-skilled labour attributed to precast concrete portal frames can be equally applied to steel portal frames. A further advantage of steel is that generally the overall dead load of a steel portal frame is less than a comparable precast concrete portal frame. However, steel has the disadvantage of being a corrosive material which will require a long life protection of a patent coating or regular protective maintenance generally by the application of coats of paint. Steel has a lower fire resistance than precast concrete but if the frame is for a single storey building structural fire protection may not be required under the Building Regulations (see Approved Document B, page 37). Typical details of steel portal frames, cladding fixings, splicing and hinges are shown in Figs. II.77 to II.79.

20
Timber portal frames

Timber portal frames can be manufactured by several methods which produce a light, strong frame of pleasing appearance which renders them suitable for buildings such as churches, halls and gymnasiums where clear space and appearance are important. The common methods used are glued laminated portal frames, plywood faced portal frames and timber portal frames using solid members connected together with plywood gussets.

GLUED LAMINATED PORTAL FRAMES

The main objective of forming a laminated member consisting of glued layers of thin section timber members is to obtain an overall increase in strength of the complete component over that which could be expected from a similar sized solid section of a particular species of timber. This type of portal frame is usually manufactured by a specialist firm since the jigs required would be too costly for small outputs. The selection of suitable quality softwoods of the right moisture content is also important for a successful design. In common with other timber portal frames, these can be fully rigid, 2 pin or 3 pin structures.

Site work is simple, consisting of connecting the foot of the supporting member to the metal shoe fixing or to a pivot housing bolted to the concrete foundation and connecting the joint at the apex or crown with a bolt fixing or a hinge device. Most glued laminated timber portal frames are fabricated in two halves which eases transportation problems and gives maximum usage of the assembly jigs. The frames can be linked together at

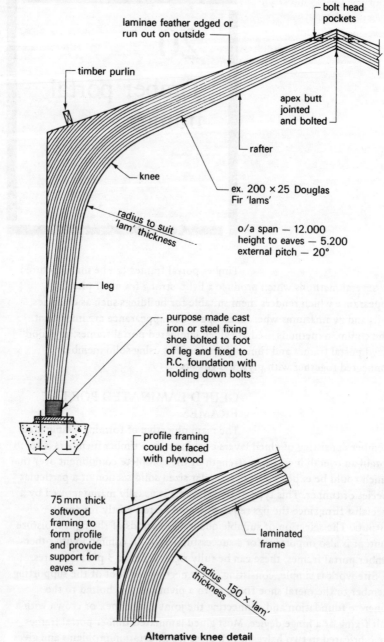

bolt head pockets

laminae feather edged or run out on outside

timber purlin

apex butt jointed and bolted

rafter

knee

ex. 200 × 25 Douglas Fir 'lams'

radius to suit 'lam' thickness

o/a span — 12.000
height to eaves — 5.200
external pitch — 20°

leg

purpose made cast iron or steel fixing shoe bolted to foot of leg and fixed to R.C. foundation with holding down bolts

profile framing could be faced with plywood

75 mm thick softwood framing to form profile and provide support for eaves

laminated frame

radius 150 × 'lam' thickness

Alternative knee detail

Fig. II.80 Typical glued laminated portal frame

roof level with timber purlins and clad with a lightweight sheeting or decking; alternatively, they may be finished with traditional roof coverings. Any form of walling can be used in conjunction with these frames provided such walling forms comply with any of the applicable Building Regulations. Typical details are shown in Fig. II.80.

PLYWOOD FACED PORTAL FRAMES

These frames are suitable for small halls, churches and schools with spans in the region of 9.000 m The portal frames are in essence boxed beams consisting of a skeleton core of softwood members faced on both sides with plywood which takes the bending stresses. The hollow form of the construction enables electrical and other small services to be accommodated within the frame members. Design concepts, fixing and finishes are as given above for glued laminated portal frames. Typical details are shown in Fig. II.81.

SOLID TIMBER AND PLYWOOD GUSSETS

These frames were developed to provide a simple and economic timber portal frame for clear span buildings using ordinary tools and basic skills. The general concept of this form of frame varies from the two types of timber portal frames previously described in that no glueing is used, the frames are spaced close together (600, 900 and 1 200 mm centres) and are clad with a plywood sheath so that the finished structure acts as a shell giving a lightweight building which is very rigid and strong. The frames can be supplied in two halves and assembled by fixing the plywood apex gussets on site before erection or alternatively they can be supplied as a complete frame ready for site erection.

The foundations for this form of timber portal frame consists of a ground beam or alternatively the frames can be fixed to the edge of a raft slab. A timber spreader or sole plate is used along the entire length of the building to receive and distribute the thrust loads of the frames. Connection to this spreader plate is made by using standard galvanised steel joists hangers or by using galvanised steel angle cleats. Standard timber windows and doors can be inserted into the side walls by trimming in the conventional way and infilling where necessary with studs, noggins and rafters. Typical details are shown in Fig. II.82.

The advantages of all timber portal frame types can be enumerated as follows:

1. Constructed from readily available materials at an economic cost.
2. Light in weight.

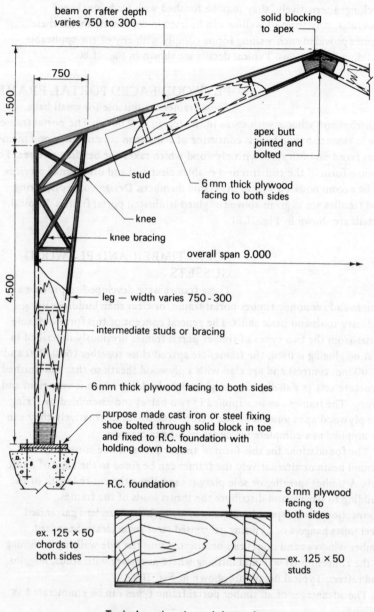

beam or rafter depth
varies 750 to 300

solid blocking
to apex

750

1.500

apex butt
jointed and
bolted

stud

6 mm thick plywood
facing to both sides

knee

knee bracing

overall span 9.000

4.500

leg — width varies 750 - 300

intermediate studs or bracing

6 mm thick plywood facing to both sides

purpose made cast iron or steel fixing
shoe bolted through solid block in toe
and fixed to R.C. foundation with
holding down bolts

R.C. foundation

6 mm plywood
facing to
both sides

ex. 125 × 50
chords to
both sides

ex. 125 × 50
studs

Typical section through leg or boom

Fig. II.81 Typical plywood faced portal frame

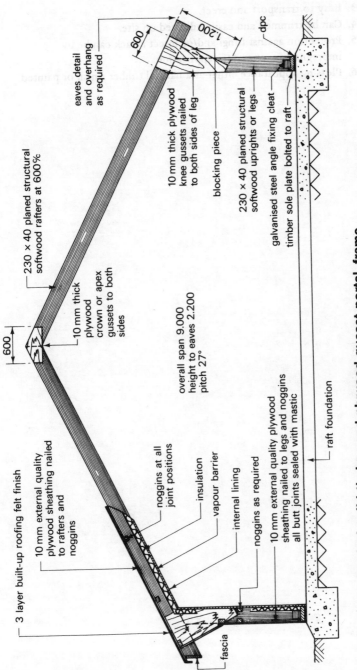

Fig. II.82 Typical solid timber and plywood gusset portal frame

eaves detail and overhang as required

dpc

600

1.200

230 × 40 planed structural softwood rafters at 600%

10 mm thick plywood knee gussets nailed to both sides of leg

blocking piece

230 × 40 planed structural softwood uprights or legs

galvanised steel angle fixing cleat

timber sole plate bolted to raft

10 mm thick plywood crown or apex gussets to both sides

600

overall span 9.000
height to eaves 2.200
pitch 27°

3 layer built-up roofing felt finish

10 mm external quality plywood sheathing nailed to rafters and noggins

noggins at all joint positions

insulation

vapour barrier

internal lining

noggins as required

10 mm external quality plywood sheathing nailed to legs and noggins all butt joints sealed with mastic

raft foundation

fascia

193

3. Easy to transport and erect.
4. Can be trimmed and easily adjusted on site.
5. Protection against fungi and/or insect attack can be by impregnation, or surface application.
6. Pleasing appearance either as a natural timber finish or painted.

21
Prestressed concrete – principles and applications

The basic principle of prestressing concrete is very simple. If a material has little tensile strength it will fracture immediately its own tensile strength is exceeded, but if such a material is given an initial compression then when load-creating tension is applied the material will be able to withstand the force of this load as long as the initial compression is not exceeded. At this stage in the study of construction technology students will already be familiar with the properties of concrete that result in a material of high compressive strength with low tensile strength and that by inserting into the member steel reinforcing bars of the correct area and fixed to a predetermined pattern ordinary concrete can be given an acceptable amount of tensile strength. Prestressing techniques are applied to concrete in an endeavour to make full use of the material's high compressive strength.

Attempts were made at the end of the nineteenth century to induce a prestress into concrete but these were largely unsuccessful since the prestress could not be maintained. A French civil engineer Marie Eugene Leon Freyssinet (1879–1962) showed in the early 1920s how this problem could be overcome and demonstrated the type of concrete and prestressing steel which was required. His most significant contributions were the quantitative assessment of creep, shrinkage and the realisation that only a high strength steel at a high stress would achieve a permanent prestress in concrete.

In normal reinforced concrete the designer is unable to make full use of the high tensile strength of steel or of the high compressive strength of the concrete. When loaded above a certain limit tension cracks will occur in a reinforced concrete member and these should not generally be greater than 0.3 mm in width as recommended in BS 8110. This stage of cracking will normally be reached before the full strength potential of both steel and concrete has been obtained. In prestressed concrete the steel is stretched and securely anchored; it will then try to regain its original length but since it is fully restricted it will be subjecting the concrete to a compressive force throughout its life. A comparison of methods is shown in Fig. II.83.

Concrete whilst curing will shrink; it will also suffer losses in cross-section due to creep when subjected to pressure. Shrinkage and creep in concrete can normally be reduced to an acceptable level by using a material of high strength with a low workability. Mild steel will also suffer from relaxation losses which is the phenomenon of the stresses in steel under load decreasing towards a minimum value after a period of time. This can be counteracted by increasing the initial stress in the steel. If mild steel is used to induce a compressive force into a concrete member the amount of shrinkage, creep and relaxation which occurs will cancel out any induced stress. The special alloy steels used in prestressing, however, have different properties enabling the designer to induce extra stress into the concrete member, thus counteracting any losses due to shrinkage and creep and at the same time maintaining the induced compressive stress in the concrete component.

The high quality strength concrete specified for prestress work should take into account the method of stressing. For pre-tensioned work a minimum 28-day cube strength of 40 N/mm^2 is required whereas for post-tensioned work a minimum 28-day cube strength of 30 N/mm^2 is required. Steel in the form of wire or bars used for prestressing should conform to the recommendations of BS 5896 which covers steel wire manufactured from cold drawn plain carbon steel. The wire can be plain round, crimped or indented with a diameter range of 2 to 7 mm. Crimped and indented bars will develop a greater bond strength than plain round bars and are available in 4, 5 and 7 mm diameters. Another form of stressing wire or tendon is strand which consists of a straight core wire around which is helically wound further wires to form a 6 over 1 or 7 wire strand or a 9 over 9 over 1 giving a 19 wire strand tendon.

Seven wire strand is the easiest to manufacture and is in general use for tendon diameters up to 15 mm. The wire used to form the strand is cold drawn from plain carbon steel as recommended in BS 5896. To ensure close contact of the individual wires in the tendon the straight core wire is usually 2% larger in diameter than the outer wires which are helically wound around it at a pitch of 12 to 16 times the nominal diameter of the strand. Nineteen wire strands with diameters ranging from 25 to 32 mm

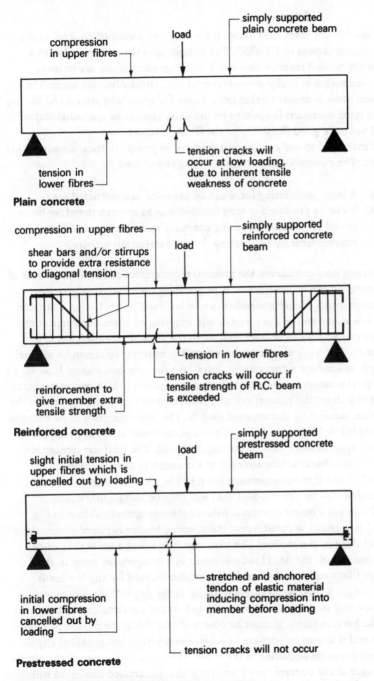

Plain concrete

- simply supported plain concrete beam
- compression in upper fibres
- load
- tension in lower fibres
- tension cracks will occur at low loading, due to inherent tensile weakness of concrete

Reinforced concrete

- simply supported reinforced concrete beam
- compression in upper fibres
- shear bars and/or stirrups to provide extra resistance to diagonal tension
- load
- tension in lower fibres
- reinforcement to give member extra tensile strength
- tension cracks will occur if tensile strength of R.C. beam is exceeded

Prestressed concrete

- simply supported prestressed concrete beam
- slight initial tension in upper fibres which is cancelled out by loading
- load
- initial compression in lower fibres cancelled out by loading
- stretched and anchored tendon of elastic material inducing compression into member before loading
- tension cracks will not occur

Fig. II.83 Structural concrete — comparison of methods

are made from wires cold drawn from patented plain carbon steel to the recommendations of BS 4757. The straight core wire is covered with a helically wound layer of nine small diameter wires which are in turn covered with a helically wound layer of nine larger diameter wires. The helical pitch is similar to that given above for seven wire strand and the use of varying diameters is again to ensure close contact of the individual wires and to create good flexibility in the finished strand.

Tendons of strand can be used singly or in groups to form a multi-strand cable. The two major advantages of using strand are:

1. A large prestressing force can be provided in a restricted area.
2. It can be produced in long flexible lengths and can therefore be stored on drums thus saving site space and reducing site labour requirements by eliminating the site fabrication activity.

Having now considered the material requirements the basic principles of prestressing can be considered. A prestressing force inducing precompression into a concrete member can be achieved by anchoring a suitable tendon at one end of the member and applying an extension force at the other end which can be anchored when the desired extension has been reached. Upon release the anchored tendon in trying to regain its original length will induce a compressive force into the member. Figure II.84 shows a typical arrangement in which the tendon inducing the compressive force is acting about the neutral axis and is stressed so that it will cancel out the tension induced by the imposed load W. The stress diagrams show that the combined or final stress will result in a compressive stress in the upper fibres equal to twice that of the imposed load. The final stress must not exceed the characteristic strength of the concrete as recommended in BS 8110 and if the arrangement given in Fig. II.84 is adopted the stress induced by the imposed load will only be half its maximum.

To obtain a better economic balance the arrangement shown in Fig. II.85 is normally adopted where the stressing tendon is placed within the lower third of the section. The basic aim is to select a stress that when combined with the dead load will result in a compressive stress in the lower fibres equal to the maximum stresses induced by any live loads, resulting in a final stress diagram having in the upper fibres a compressive stress equal to the characteristic strength of the concrete and a zero stress in the bottom fibres. It must be observed that this is the pure theoretical case and is almost impossible to achieve in practice, but provided any induced tension occurring in the lower fibres is not in excess of the tensile strength of the concrete used, an acceptable prestressed condition will exist.

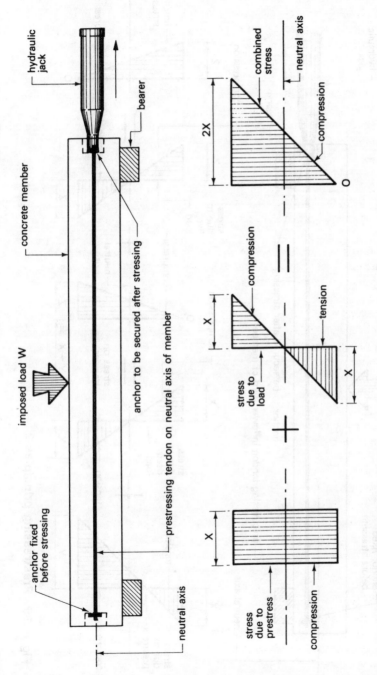

Fig. II.84 Prestressing principles 1

200

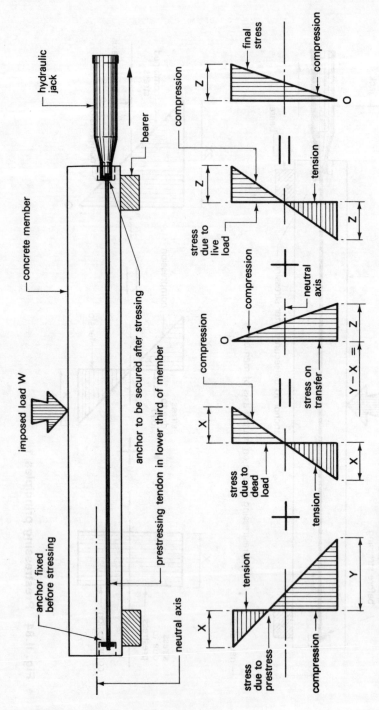

Fig. II.85 Prestressing principles 2

PRESTRESSING METHODS

There are two methods of producing prestressed concrete members namely:

1. Pre-tensioning.
2. Post-tensioning.

Pre-tensioning: in this method the wires or cables are stressed before the concrete is cast around them. The stressing wires are anchored at one end of the mould and stressed by hydraulic jacks from the other end until the required stress has been obtained. It is common practice to overstress the wires by some 10% to counteract the anticipated losses which will occur due to creep, shrinkage and relaxation. After stressing the wires the side forms of the mould are positioned and the concrete is placed around the tensioned wires; the casting is then usually steam cured for 24 hours to obtain the desired characteristic strength, a common specification being 28 N/mm^2 in 24 hours. The wires are cut or released and the bond between the stressed wires and concrete will prevent the tendons from regaining their original length thus inducing the prestress.

At the extreme ends of the members the bond between the stressed wires and concrete will not be fully developed owing to low frictional resistance resulting in a contraction and swelling at the ends of the wires forming what is in effect a cone shape anchor. The distance over which this contraction takes place is called the transfer length and is equal to 80 to 120 times the wire diameter. Usually small diameter wires (2 to 5 mm) are used so that for any given total area of stressing wire a greater surface contact area is obtained. The bond between the stressed wires and concrete can also be improved by using crimped or indented wires.

Pre-tensioning is the prestressing method used mainly by manufacturers of precast components such as floor units and slabs employing the long line method of casting where precision metal moulds up to 120.000 long can be used with spacers or dividing plates positioned along the length to create the various lengths required — a typical arrangement is shown in Fig. II.86.

Post-tensioning: in this method the concrete is cast around ducts in which the stressing tendons can be housed and the stressing is carried out after the concrete has hardened. The tendons are stressed from one or both ends and when the stress required has been reached the tendons are anchored at their ends to prevent them returning to their original length thus inducing the compressive force. The anchors used form part of the finished component. The ducts for housing the stressing tendons can be formed by using flexible steel tubing or inflatable rubber tubes. The void created by

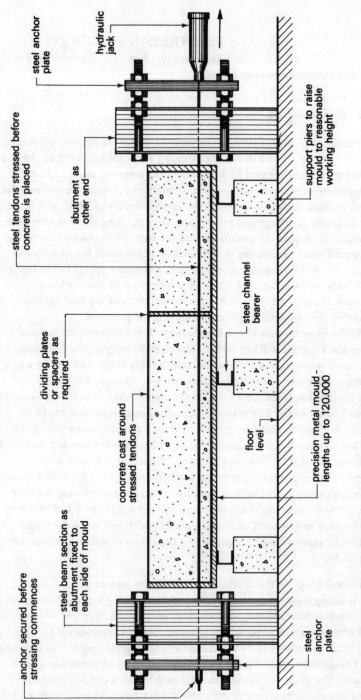

Fig. II.86 Typical pre-tensioning arrangement

hydraulic jack

steel anchor plate

support piers to raise mould to reasonable working height

steel tendons stressed before concrete is placed

abutment as other end

steel channel bearer

dividing plates or spacers as required

floor level

precision metal mould - lengths up to 120.000

concrete cast around stressed tendons

steel beam section as abutment fixed to each side of mould

steel anchor plate

anchor secured before stressing commences

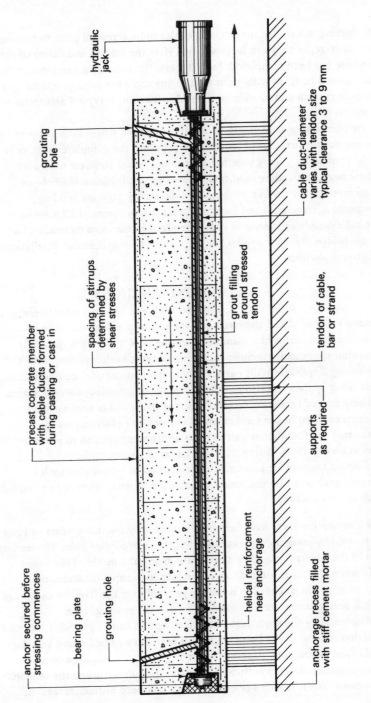

hydraulic jack

grouting hole

cable duct-diameter varies with tendon size typical clearance 3 to 9 mm

precast concrete member with cable ducts formed during casting or cast in

spacing of stirrups determined by shear stresses

grout filling around stressed tendon

tendon of cable, bar or strand

supports as required

anchor secured before stressing commences

bearing plate

grouting hole

helical reinforcement near anchorage

anchorage recess filled with stiff cement mortar

Fig. II.87 Typical post-tensioning arrangement

203

the ducting will enable the stressing cables to be threaded prior to placing the concrete, or they can be positioned after the casting and curing of the concrete has been completed. In both cases the remaining space within the duct should be filled with grout to stop any moisture present setting up a corrosive action and to assist in stress distribution. A typical arrangement is shown in Fig. II.87.

Post-tensioning is the method usually employed where stressing is to be carried out on site, curved tendons are required, the complete member is to be formed by joining together a series of precast concrete units and where negative bending moments are encountered. Figure II.88 shows diagrammatically various methods of overcoming negative bending moments at fixed ends and for continuous spans. Figure II.89 shows a typical example of the use of curved tendons in the cross members of a girder bridge. Another application of post-tensioning is in the installation of ground anchors.

GROUND ANCHORS

A ground anchor is basically a prestressing tendon embedded and anchored into the soil to provide resistance to structural movement of a member by acting on a 'tying back' principle. Common applications include anchoring or tying back retaining walls and anchoring diaphragm walls particularly in the context of deep excavations. This latter application also has the advantage of providing a working area entirely free of timbering members such as struts and braces. Ground anchors can also be used in basement and similar constructions for anchoring the foundation slab to resist uplift pressures and to prevent floatation especially during the early stages of construction.

Ground anchors are known by their method of installation such as grouted anchors or by the nature of subsoil into which they are embedded such as rock anchors.

Rock anchors — these have been used successfully for many years and can be formed by inserting a prestressing bar into a predrilled hole. The leading end of the bar has expanding sleeves which grip the inside of the bored hole when the bar is rotated to a recommended torque to obtain the desired grip. The anchor bar is usually grouted over the fixed or anchorage length before being stressed and anchored at the external face.

Alternatively the anchorage of the leading end can be provided by grout injection relying on the bond developed between a ribbed sleeve and the wall of the bored hole. A dense high strength grout is required over the fixed length to develop sufficient resistance to pull out when the tendon is stressed. The unbonded or elastic length will need protection against

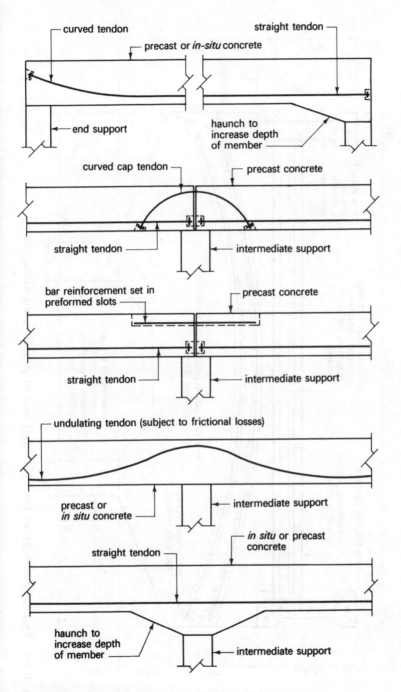

Fig. II.88 Prestressing — overcoming negative bending moments

205

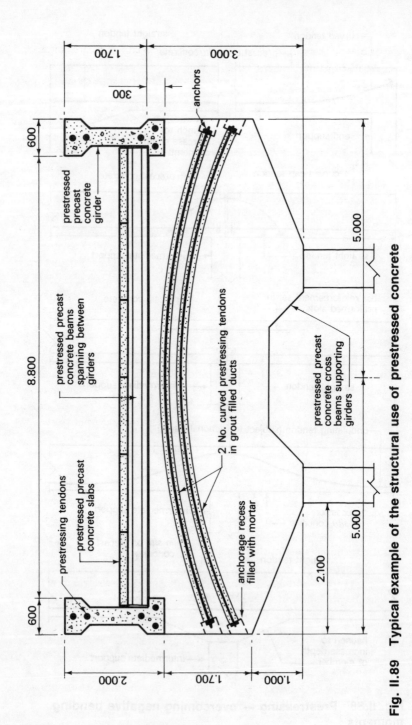

600

1.700

300

3.000

anchors

600

prestressed precast concrete girder

8.800

prestressed precast concrete beams spanning between girders

5.000

prestressing tendons

prestressed precast concrete slabs

2 No. curved prestressing tendons in grout filled ducts

prestressed precast concrete cross beams supporting girders

anchorage recess filled with mortar

5.000

2.000

1.700

1.000

2.100

Fig. II.89 Typical example of the structural use of prestressed concrete

206

corrosion and this can be provided by protective coatings such as bitumen or rubberised paint, casings of PVC, wrappings of greased tape or a full-length protection can be given by filling the void with grout after completion of the stressing operation — see Fig. II.90.

Injection anchors — the knowledge and experience gained in the use of rock anchors has led to the development of suitable ground anchorage techniques for most subsoil conditions except for highly compressible soils such as alluvial clays and silts. Injection-type ground anchorages have proved to be suitable for most cohesive and non-cohesive soils. Basically a hole is bored into the soil using a flight auger with or without water flushing assistance; casings or linings can be used where the borehole would not remain open if unlined. The prestressing tendon or bar is placed into the borehole and pressure grouted over the anchorage length. For protection purposes the unbonded or elastic length can be grouted under gravity for permanent ground anchors or covered with an expanded polypropylene sheath for temporary anchors. Anchor boreholes in clay soils are usually multi-under-reamed to increase the bond, using special expanding cutter or brush tools. Gravel placement ground anchors can also be used in clay and similar soils for lighter loadings. In this method an irregular gravel is injected into the borehole over the anchorage length. A small casing with a non-recoverable point is driven into the gravel plug to force the aggregate to penetrate the soil around the borehole. The stressing tendon is inserted into the casing and pressure grouted over the anchorage length as the casing is removed. Typical ground anchor examples are shown in Fig. II.91.

The calculation of anchorage lengths, number of anchors, spacing of anchors and tendon stressing requirements are the province of the engineer and are therefore not considered in this text.

The above principles and applications make it clear that in pre-tensioning it is the bond between the tendon and concrete which prevents the prestressing wire returning to its original length and in post-tensioning it is the anchorages which prevent the stressing tendon returning to its original length.

The advantages and disadvantages of prestressed concrete when compared with conventional reinforced concrete can be enumerated thus:

Advantages
1. Makes full use of the inherent compressive strength of concrete.
2. Makes full use of the special alloy steels used to form prestressing tendons.
3. Eliminates tension cracks thus reducing the risk of corrosion of steel components.

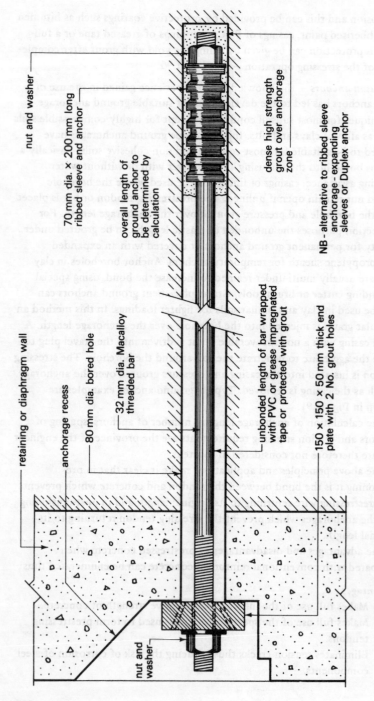

Fig. II.90 Typical rock ground anchor

nut and washer

dense high strength
grout over anchorage
zone

70 mm dia. × 200 mm long
ribbed sleeve and anchor

overall length of
ground anchor to
be determined by
calculation

NB - alternative to ribbed sleeve
anchorage - expanding
sleeves or Duplex anchor

retaining or diaphragm wall

anchorage recess

80 mm dia. bored hole

32 mm dia. Macalloy
threaded bar

unbonded length of bar wrapped
with PVC or grease impregnated
tape or protected with grout

150 × 150 × 50 mm thick end
plate with 2 No. grout holes

nut and
washer

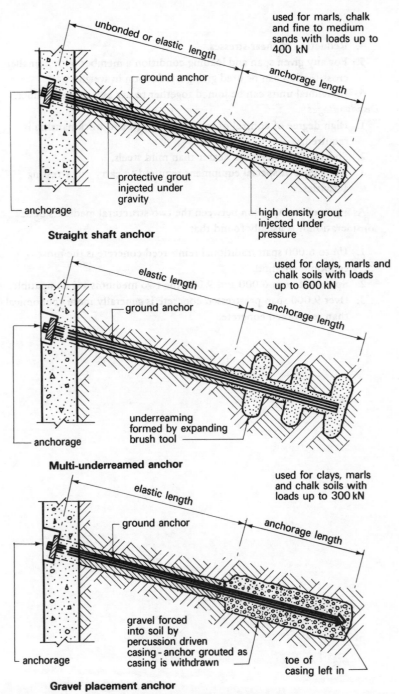

Fig. II.91 Typical injection ground anchors

209

4. Reduction in shear stresses.
5. For any given span and loading condition a member with a smaller cross-section can be used giving a reduction in weight.
6. Individual units can be joined together to act as a single member.

Disadvantages
1. High degree of control of materials, design and workmanship is required.
2. Special alloy steels are dearer than mild steels.
3. Extra cost of special equipment required to carry out stressing activities.

As a general comparison between the two structural mediums under consideration it is usually found that:

1. Up to 6.000 span traditional reinforced concrete is the most economic method.
2. Spans between 6.000 and 9.000 the two mediums are compatible.
3. Over 9.000 span prestressed concrete is generally more economical than reinforced concrete.

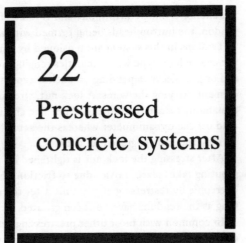

22
Prestressed concrete systems

The prestressing of concrete is usually carried out by a specialist contractor or alternatively by the main contractor using a particular system and equipment. The basic conception and principles of prestressing are common to all systems; it is only the type of tendon, type of anchorage and stressing equipment which varies. The following systems are typical and representative of the methods available.

BBRV (Simon-Carves Ltd)

This is a unique system of prestressing developed by four Swiss engineers namely Birkenmaier, Brandestini, Ros and Vogt whose initials are used to name the system. It differs from other systems in that multi-wire cables capable of providing prestressing forces from 300 to 7 800 kN are used and each wire is anchored at each end by means of enlarged heads formed on the wire. The cables or tendons are purpose made to suit individual requirements and may comprise up to 121 wires. The high tensile steel wire conforming to the recommendations of BS 5896 is cut to the correct length, any sheathing required is threaded on together with the correct anchorage before the button heads are formed using special equipment. The completed tendons can be coiled or left straight for delivery to site. Four types of stressing anchor are available, the choice being dependent on the prestressing force being induced, whereas the fixed anchors come in three forms. The finished tendon is fixed in the correct position to the formwork before concreting; when the concrete has hardened the tendons are stressed and grout is injected into the sheathing. If unsheathed tendons are to be drawn into preformed ducts the anchor is omitted from one and is fixed after drawing through the tendon, the button heads being formed with a portable machine.

Tendons in this system are tensioned by using a special hydraulic jack of the centre hole type with capacities ranging from 30 to 800 tonnes. A pull rod or pull sleeve depending on anchor type is coupled to the basic element carrying the wires. A lock nut, stressing stool, hydraulic jack and dynamometer are then threaded on. The applied stressing force can be read off the dynamometer whereas the actual extension achieved can be seen by the scale engraved on the jack.

After stressing the lock nut is tightened up and the jack released before grouting takes place. Losses due to friction, shrinkage and creep can be overcome by restressing at any time after the initial stressing operation so long as the tendons have not been grouted.

In common with most other prestressing systems tendons in long continuous members can be stressed in stages without breaking the continuity of the design. The first tendon length is stressed and grouted before the second tendon length is connected to it using a coupling anchor, after which it is stressed and grouted before repeating the procedure for any subsequent lengths. Alternatively, lengths of unstressed tendons can be coupled together and the complete tendon stressed from one or both ends. Figure II.92 shows a typical BBRV stressing arrangement.

CCL (CCL Systems Ltd)

This post-tensioning system uses a number of strands to form the tendon ranging from 4 to 31 strands according to the system being employed giving a range of prestressing forces from 450 to 5 000 kN. Two basic systems are available namely Cabco and Multiforce. In the Cabco system each strand or wire is stressed individually, the choice of hydraulic jack being governed by the size of strand being used. The system is fast and being manually operated eliminates the need for lifting equipment. By applying the total tendon force in stages problems such as differential elastic shortening and out of balance forces are reduced. Curved tendons are possible using this system without the need for spacers but spacers are recommended for tendons over 30.000 long.

The alternative Multiforce system uses the technique of simultaneously stressing all the strands forming the tendon. In both systems the basic anchorages are similar in design. The fixed or dead-end anchorage has a tube unit to distribute the load, a bearing plate and a compressing grip fitted to each strand. If the fixed anchorage is to be totally embedded in concrete a compressible gasket together with a bolted-on retaining plate are also used, a grout vent pipe being inserted into the grout hole of the tube unit. The compression grips consist of an outer sleeve with a

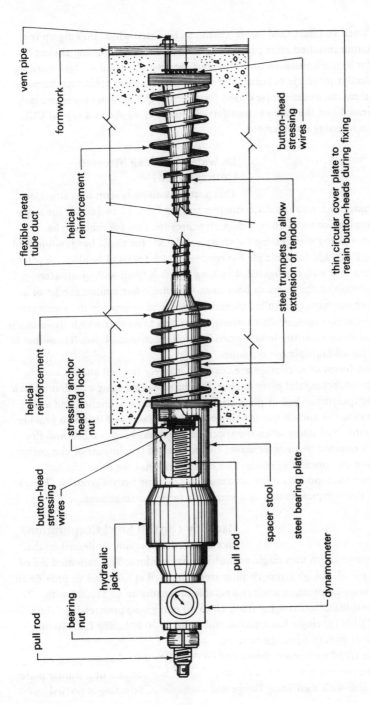

Fig. II.92 Typical BBRV prestressing arrangement

Labels (clockwise from top):
- vent pipe
- formwork
- button-head stressing wires
- flexible metal tube duct
- helical reinforcement
- thin circular cover plate to retain button-heads during fixing
- steel trumpets to allow extension of tendon
- helical reinforcement
- stressing anchor head and lock nut
- button-head stressing wires
- spacer stool
- steel bearing plate
- dynamometer
- pull rod
- hydraulic jack
- bearing nut
- pull rod

machined and hardened insert which can be fitted whilst making up the tendon or installed after positioning the tendon. The stressing anchor is a similar device consisting of a tube unit, bearing plate and wedges working on a collet principle to secure the strands. The wedges are driven home to a 'set' by the hydraulic jack used for stressing the tendon before the jack is released and the stress is transferred. Figure II.93 shows a typical CCL Cabco stressing arrangement.

Dywidag (Dividag Stressed Concrete Ltd)

This post-tensioning system uses single or multiple bar tendons with diameters ranging from 12 to 36 mm for single bar applications and 16 mm diameter threaded bars for multiple bar tendons giving prestressing forces up to 950 kN for single bar tendons and up to 2 000 kN for multiple bar tendons. Both forms of tendon are placed inside a thin wall corrugated sheathing which is filled with grout after completion of the stressing operation. The single bar tendon can be of a smooth bar with cold rolled threads at each end to provide the connection for the anchorages, or alternatively it can be a threadbar which has a coarse thread along its entire length providing full mechanical bond. Threadbar is used for all multiple bar tendons.

Two forms of anchorage are available, namely the bell anchor and the square or rectangular plate anchor. During stressing, using a hydraulic jack acting against the bell or plate anchor, the tendon is stretched and at the same time the anchor nut is being continuously screwed down to provide the transfer of stress when the specified stress has been reached and the jack is released. A counter shows the number of revolutions of the anchor nut and the amount of elongation of the tendon. Like other similar systems the tendons can be restressed at any time before grouting. Figure II.94 shows typical details of a single bar tendon arrangement.

Macalloy (British Steel Corporation)

Like the method previously described the Macalloy system uses single or multiple bar tendons. The bars used are of a cold worked high strength alloy steel threaded at each end to provide an anchorage connection and are available in lengths up to 18.000 with nominal diameters ranging from 20 to 40 mm giving prestressing forces up to 875 kN for single bar tendons and up to 3500 kN using four 40 mm diameter bars to form the tendon.

The fixed anchorage consists of an end plate drilled and tapped to receive the tendon whereas the stressing anchor consists of a similar plate complete with a grouting flange and anchor nut. Stressing is carried out

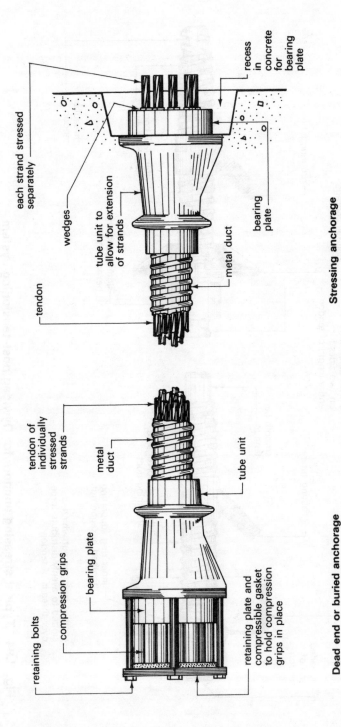

each strand stressed
separately

recess
in
concrete
for
bearing
plate

wedges

tube unit to
allow for extension
of strands

tendon

metal duct

bearing
plate

Stressing anchorage

tendon of
individually
stressed
strands

metal
duct

tube unit

retaining bolts

compression grips

bearing plate

retaining plate and
compressible gasket
to hold compression
grips in place

Dead end or buried anchorage

Fig. II.93   Typical details of the CCL Cabco system of prestressing

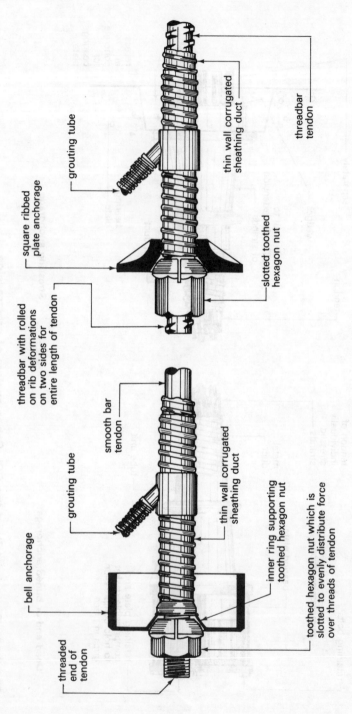

grouting tube

square ribbed plate anchorage

thin wall corrugated sheathing duct

threadbar tendon

slotted toothed hexagon nut

threadbar with rolled on rib deformations on two sides for entire length of tendon

smooth bar tendon

grouting tube

bell anchorage

thin wall corrugated sheathing duct

inner ring supporting toothed hexagon nut

threaded end of tendon

toothed hexagon nut which is slotted to evenly distribute force over threads of tendon

Fig. II.94 Typical stressing anchors for Dywidag post-tensioning system

using a hydraulic jack operating on a drawbar attached to the tendon, the tightened anchor nut transferring the stress to the member upon completion of the stressing operation. The prestressing is completed by grouting in the tendon after all the stressing and any necessary restressing has been completed. Figure II.95 shows typical details of a single bar arrangement.

PSC (PSC Equipment Ltd)

This post-tension system uses strand to form the tendon and is available in three forms, namely the Freyssi-monogroup, Freyssinet multistrand and PSC monostrand. Monogroup tendons are composed of 7, 13, 15 or 19 wire strands stressed in a single pull by the correct model of hydraulic jack giving prestressing forces of up to 5 000 kN. The anchorages consist of a cast iron guide shaped to permit the deviation of the strands to their positions in the steel anchor block where they are secured by collet-type jaws. During stressing 12 wires of the tendon are anchored to the body of the jack, the remaining wires passing through the jack body to an anchorage at the rear.

The multistrand system is based upon the first system of post-tension ever devised and consists of a cable tendon made from 12 high-tensile steel wires laid parallel to one another and taped together resulting in a tendon which is flexible and compact. Two standard cable diameters are produced giving tendon sizes of 29 and 33 mm capable of taking prestressing forces up to 750 kN. The anchorages consist of two parts, the outer reinforced concrete cylindrical body with a tapered hole to receive a conical wedge which is grooved or fluted to receive the wires of the tendon. The 12 wires of the cable are wedged into tapered slots on the outside of the hydraulic jack body during stressing and when this operation has been completed the jack drives home the conical wedge to complete the anchorage.

Monostrand uses a 4 or 7 strand tendon for general prestressing or a 3 strand tendon developed especially for prestressing floor and roof slabs. This system is intended for the small to medium range of prestressing work requiring a prestressing force not exceeding 2 000 kN and as its title indicates each strand in the group is stressed separately requiring only a light compact hydraulic jack. All PSC system tendons are encased in a steel sheath and grouted after completion of the stressing operation. Typical details are shown in Fig. II.96.

SCD (Stressed Concrete Design Ltd)

This post-tensioning system offers three variations of tendon and/or stressing, namely multigrip circular, monogrip circular and monogrip rectangular. The multigrip circular system uses a tendon of 7, 12, 13 or 19 strands forming a cable tendon capable of

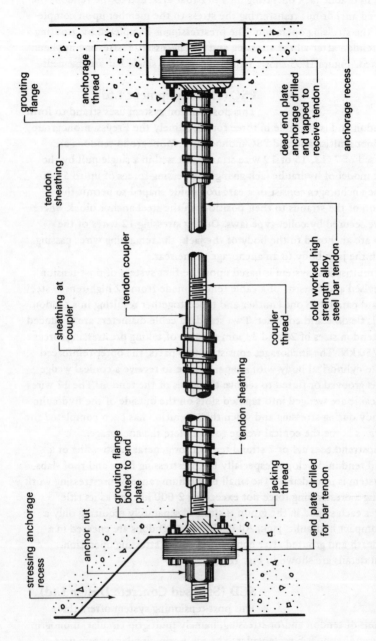

Fig. II.95 Typical Macalloy prestressing system details

grouting flange

anchorage thread

dead end plate anchorage drilled and tapped to receive tendon

anchorage recess

tendon sheathing

sheathing at coupler

tendon coupler

coupler thread

cold worked high strength alloy steel bar tendon

tendon sheathing

grouting flange bolted to end plate

jacking thread

stressing anchorage recess

anchor nut

end plate drilled for bar tendon

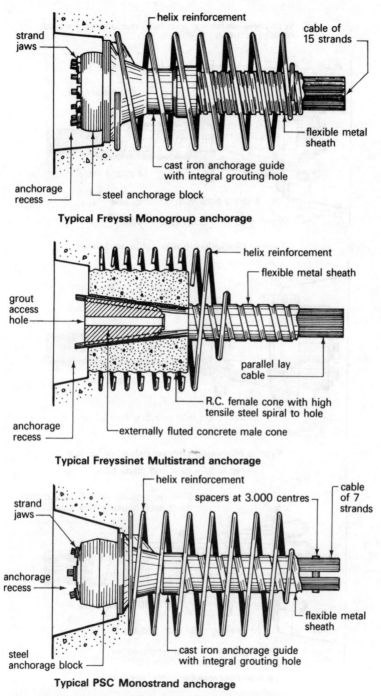

Typical Freyssi Monogroup anchorage

Typical Freyssinet Multistrand anchorage

Typical PSC Monostrand anchorage

Fig. II.96 Typical PSC prestressing system

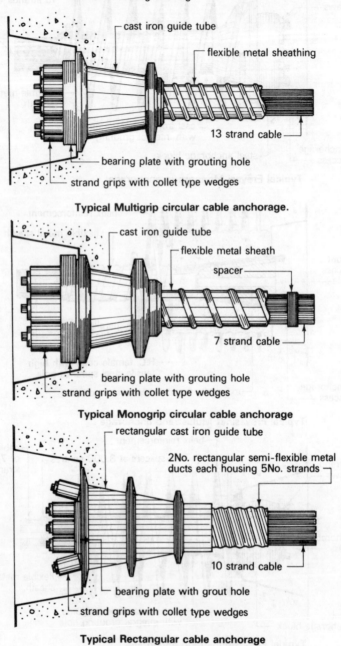

NB. in all cases anchorage zone helix reinforcement would be used according to design

cast iron guide tube

flexible metal sheathing

13 strand cable

bearing plate with grouting hole

strand grips with collet type wedges

Typical Multigrip circular cable anchorage.

cast iron guide tube

flexible metal sheath

spacer

7 strand cable

bearing plate with grouting hole

strand grips with collet type wedges

Typical Monogrip circular cable anchorage

rectangular cast iron guide tube

2No. rectangular semi-flexible metal ducts each housing 5No. strands

10 strand cable

bearing plate with grout hole

strand grips with collet type wedges

Typical Rectangular cable anchorage

Fig. II.97 Typical SCD prestressing system details

220

accepting prestressing forces up to 5 000 kN. The anchorages consist of a cast iron guide plate enabling the individual strands of the cable to fan out and pass through a bearing plate where they are secured with steel collet-type wedges. The strands pass through the body of the hydraulic jack to a rear anchorage and are therefore stressed simultaneously.

The monogrip circular tendons are available as a single, 4, 7 or 12 strand cable in which each strand is stressed individually and the whole tendon being capable of taking prestressing forces up to 3 200 kN. Each strand in the tendon is separated from adjacent strands by means of circular spacers at 2.000 centres or less if the tendon is curved. The anchorages are similar in principle to those already described for multi-grip tendons.

The monogrip rectangular system by virtue of using a rectangular tendon composed of 3 to 27 strands capable of taking a prestressing force of up to 3 900 kN affords maximum eccentricity with a wide range of tendon sizes. The anchorage guide plate and bearing plates work on the same principle as described above for the multigrip method. In all methods the tendon is encased in a sheath unless preformed ducts have been cast; upon completion of the stressing operation all tendons are grouted in. See Fig. II.97 for typical details.

Part III
Floors

23
Flooring – solid ground, suspended timber

SOLID GROUND FLOORS

The construction of a solid ground floor can be considered under three headings:

1. Hardcore.
2. Blinding.
3. Concrete bed or slab.

Hardcore

The purpose of hardcore is to fill in any small pockets which have formed during oversite excavations, to provide a firm base on which to place a concrete bed and to help spread any point loads over a greater area. It also acts against capillary action of moisture within the soil. Hardcore is usually laid in 100-150 mm layers to the required depth and it is important that each layer is well compacted, using a roller if necessary, to prevent any unacceptable settlement beneath the solid floor.

Approved Document C recommends that no hardcore laid under a solid ground floor shall contain water-soluble sulphates or other harmful matter in such quantities as to be liable to cause damage to any part of the floor. This recommendation prevents the use of any material which may swell upon becoming moist, such as colliery shale, and furthermore it is necessary to ascertain that brick rubble from demolition works and clinker furnace waste intended for use as hardcore does not have any harmful water-soluble sulphate content.

Blinding

This is used to even off the surface of hardcore if a damp-proof membrane is to be placed under the concrete bed or if a reinforced concrete bed is specified. Firstly, it will prevent the damp-proof membrane from being punctured by the hardcore and, secondly, it will provide a true surface from which the reinforcement can be positioned. Blinding generally consists of a layer of fine ash or sand 25-50 mm thick or a 50-75 mm layer of weak concrete (1 : 12 mix usually suitable) if a true surface for reinforced concrete is required.

Concrete bed

Thicknesses generally specified are:

1. Unreinforced or plain *in situ* concrete, 100-150 mm thick.
2. Reinforced concrete, 150 mm minimum.

Suitable concrete mixes are:

(a) Plain *in situ* concrete, 1 : 3 : 6 or 1 : 6 'all-in'.
(b) Reinforced concrete, 1 : 2 : 4.

The reinforcement used in concrete beds for domestic work is usually in the form of a welded steel fabric to BS 4483. Sometimes a light square mesh fabric is placed 25 mm from the upper surface of the concrete bed to prevent surface crazing and limit the size of any cracking.

In domestic work the areas of concrete are defined by the room sizes and it is not usually necessary to include expansion or contraction joints in the construction of the bed.

PROTECTION OF FLOORS NEXT TO THE GROUND

Building Regulation C4 requires that such part of a building as is next to the ground shall have a floor so constructed as to prevent the passage of moisture from the ground to the upper surface of the floor. The requirements of this regulation can only be properly satisfied by the provision of a suitable barrier in the form of a damp-proof membrane within the floor. The membrane should be turned up at the edges to meet and blend with the damp-proof course in the walls to prevent any penetration of moisture by capillary action at edges of the bed.

226

Suitable materials for damp-proof membranes are:

1. Waterproof building papers—conforming to BS 1521.
2. Polythene sheet 1 000 gauge sheet with sealed joints is acceptable and will also give protection against moisture vapour as well as moisture.
3. Hot poured bitumen—should be at least 3 mm thick.
4. Cold applied bitumen/rubber emulsions—should be applied in not less than three coats.
5. Asphalt—could be dual purpose finish and damp-proof membrane.

The position of a damp-proof membrane, whether above or below the concrete bed, is a matter of individual choice. A membrane placed above the bed is the easiest method from a practical aspect and is therefore generally used. A membrane placed below the bed has two advantages: firstly, it will keep the concrete bed dry and in so doing will make the bed a better thermal insulator and, secondly, during construction it will act as a separating layer preventing leakage of the cement matrix into the hardcore layer which could result in a weak concrete mix. Typical details of solid floor construction are shown in **Figs. III.1 and III.2.**

SUSPENDED TIMBER GROUND FLOORS

This type of floor consists of timber boards or other suitable sheet material fixed to joists spanning over sleeper walls and was until 1939 the common method of forming ground floors in domestic buildings. The Second World War restricted the availability of suitable timber and solid ground floors replaced suspended timber floors. Today the timber floor is still used on occasions because it has some flexibility and will easily accept nail fixings—properties which a solid ground floor lacks. It is a more expensive form of construction than a solid floor and can only be justified on sloping sites which would need a great deal of filling to make up the ground to the required floor level.

Suspended timber ground floors are susceptible to dry rot, draughts and are said to be colder than other forms of flooring. If the floor is correctly designed and constructed these faults can be eliminated.

The problem of dry rot, which is a fungus that attacks damp timber, can be overcome by adequate ventilation under the floor and the correct positioning of damp-proof courses to keep the under floor area and timber dry. Through ventilation is essential to keep the moisture content of the timber below that which would allow fungal growth to take place; that is, 20% of its oven-dry weight. The usual method is to allow a free flow of air under the floor covering by providing, in the external walls, air bricks sited near the corners and at approximately 2 000 mm centres

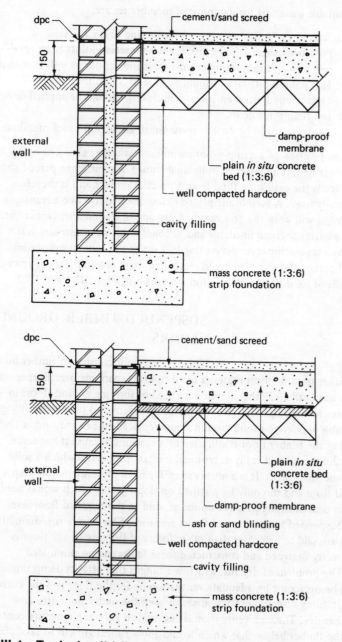

Fig. III.1 Typical solid floor details at external walls

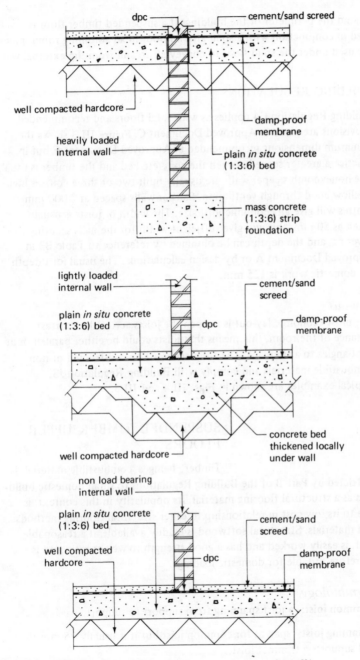

dpc — cement/sand screed

well compacted hardcore

damp-proof membrane

heavily loaded internal wall

plain *in situ* concrete (1:3:6) bed

mass concrete (1:3:6) strip foundation

lightly loaded internal wall

cement/sand screed

plain *in situ* concrete (1:3:6) bed

dpc

damp-proof membrane

well compacted hardcore

concrete bed thickened locally under wall

non load bearing internal wall

plain *in situ* concrete (1:3:6) bed

cement/sand screed

well compacted hardcore

dpc

damp-proof membrane

Fig. III.2 Typical solid floor details at internal walls

around the perimeter of the building. If a suspended timber floor is used in conjunction with a solid ground floor in an adjoining room, pipes are used under the solid floor to convey air to and from the external walls.

BUILDING REGULATIONS

Building Regulation C4 applies as with solid floors and recommended provisions are given in Approved Document C. Figure III.3 shows the minimum dimensions recommended in Approved Document C but in practice a greater space between the concrete bed and the timber is usual. The honeycomb sleeper walls are usually built two or three courses high to allow good through ventilation. Sleeper walls spaced at 2 000 mm centres will give an economic joist size. The width of joists is usually taken as 50 mm, this will give sufficient width for the nails securing the covering, and the depth can be obtained by reference to Table B3 in Approved Document A or by design calculations. The usual joist depth for domestic work is 125 mm.

Lay-out

The most economic lay-out is to span the joists across the shortest distance of the room, this means that joists could be either parallel or at right-angles to a fireplace. The fireplace must be constructed of non-combustible materials and comply with Building Regulation J3. Typical examples are shown in Figs. III.4 and III.5.

SUSPENDED TIMBER UPPER FLOORS

Timber, being a combustible material, is restricted by Part B of the Building Regulations to small domestic buildings as a structural flooring material. Its popularity in this context, is due to its low cost in relationship to other structural flooring methods and materials. Structural softwood is readily available at a reasonable cost, is easily worked and has a good strength to weight ratio and is therefore suitable for domestic loadings.

Terminology

Common joist: a joist spanning from support to support.

Trimming joist: span as for common joist but it is usually 25 mm thicker and supports a trimmer joist.

Trimmer joist: a joist at right-angles to the main span supporting the trimmed joists and is usually 25 mm thicker than a common joist.

230

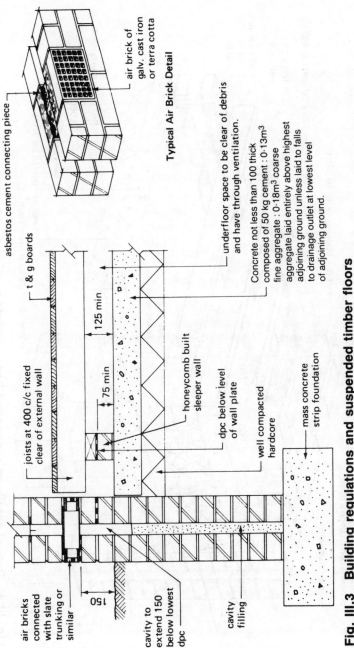

air brick of galv. cast iron or terra cotta

asbestos cement connecting piece

Typical Air Brick Detail

t & g boards

joists at 400 c/c fixed clear of external wall

125 min

75 min

honeycomb built sleeper wall

dpc below level of wall plate

well compacted hardcore

mass concrete strip foundation

underfloor space to be clear of debris and have through ventilation.

Concrete not less than 100 thick composed of 50 kg cement : 0·13m³ fine aggregate : 0·18m³ coarse aggregate laid entirely above highest adjoining ground unless laid to falls to drainage outlet at lowest level of adjoining ground.

air bricks connected with slate trunking or similar

150

cavity to extend 150 below lowest dpc

cavity filling

Fig. III.3 Building regulations and suspended timber floors

231

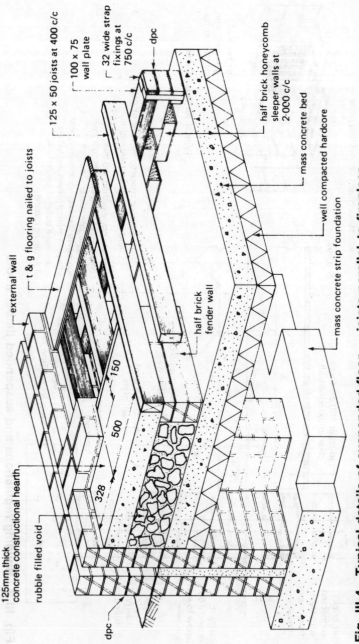

125mm thick concrete constructional hearth

rubble filled void

external wall

t & g flooring nailed to joists

125 × 50 joists at 400 c/c

100 × 75 wall plate

32 wide strap fixings at 750 c/c

dpc

half brick honeycomb sleeper walls at 2·000 c/c

mass concrete bed

well compacted hardcore

mass concrete strip foundation

half brick fender wall

dpc

150

500

328

Fig. III.4 Typical details of suspended floor — joists parallel to fireplace

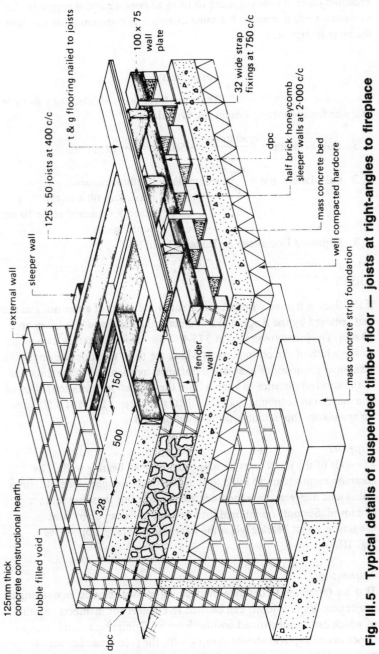

100 × 75 wall plate

t & g flooring nailed to joists

125 × 50 joists at 400 c/c

32 wide strap fixings at 750 c/c

dpc

half brick honeycomb sleeper walls at 2·000 c/c

mass concrete bed

well compacted hardcore

external wall

sleeper wall

fender wall

150

500

328

mass concrete strip foundation

125mm thick concrete constructional hearth

rubble filled void

dpc

Fig. III.5 Typical details of suspended timber floor — joists at right-angles to fireplace

233

Trimmed joist: a joist cut short to form an opening and is supported by a trimmer joist; it spans in the same direction as common joists and is of the same section size.

Joist Sizing

There are three ways of selecting a suitable joist size for supporting a domestic type floor:

1. Rule of thumb: $\dfrac{\text{span in mm}}{24} + 50 \text{ mm} = \text{depth in mm}.$

2. Calculation: $BM = \dfrac{fbd^2}{6}$ where BM = bending moment
   f = maximum fibre stress
   b = breadth (assumed to be 50 mm)
   d = depth in mm.

3. Approved Document A, Tables B3 and B4.

JOISTS

If the floor is framed with structural softwood joists of a size not less than that required by the Approved Document, the usual width is taken as 50 mm. The joists are spaced at 375-450 mm centre to centre depending on the width of the ceiling boards which are to be fixed on the underside. Maximum economy of joist size is obtained by spanning in the direction of the shortest distance to keep within the deflection limitations allowed. The maximum economic span for joists is between 3 500 and 4 500 mm, for spans over this a double floor could be used.

Support

The ends of the joists must be supported by load bearing walls. The common methods are to build in the ends or to use special metal fixings called joist hangers; other methods are possible but these are seldom employed. Support on internal load bearing walls can be by joist hangers or direct bearing when the joists are generally lapped (see Fig. III.6).

Trimming

This is a term used to describe the framing of joists around an opening or projection. Various joints can be used to connect the members together, all of which can be substituted by joist hangers. Trimming around flues and upper floor fireplaces should comply with the recommendations of Approved Document J. It should be noted that, since central heating is becoming commonplace, the provision of upper floor fireplaces is seldom

234

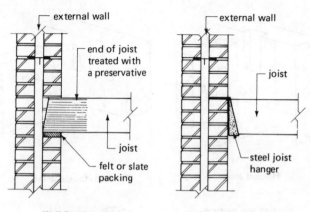

Wall Bearing

Joist Hanger Bearing

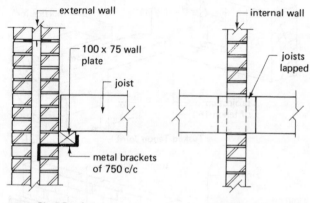

Plate Bearing

Direct Bearing

Typical Joist Hangers

Fig. III.6 Typical joist support details

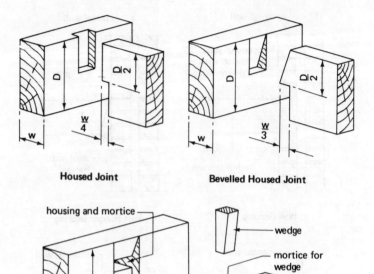

Housed Joint

Bevelled Housed Joint

housing and mortice

wedge

mortice for wedge

D

projection of tenon 150

Tusked Tenon Joint

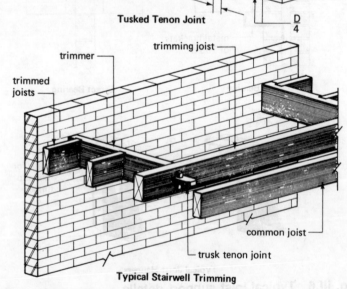

trimmed joists

trimmer

trimming joist

common joist

trusk tenon joint

Typical Stairwell Trimming

Fig. III.7 Floor trimming joints and details

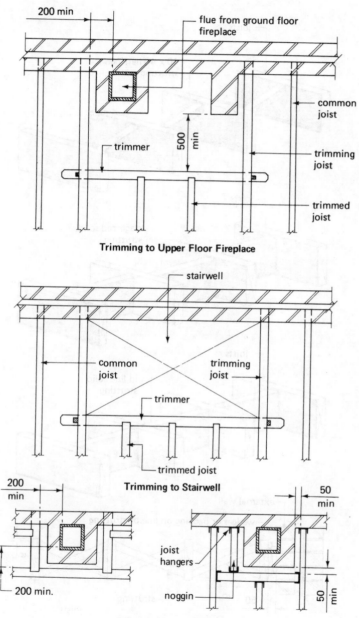

Trimming to Upper Floor Fireplace

Trimming to Stairwell

Trimming Around Flues

Fig. III.8 Typical trimming arrangements

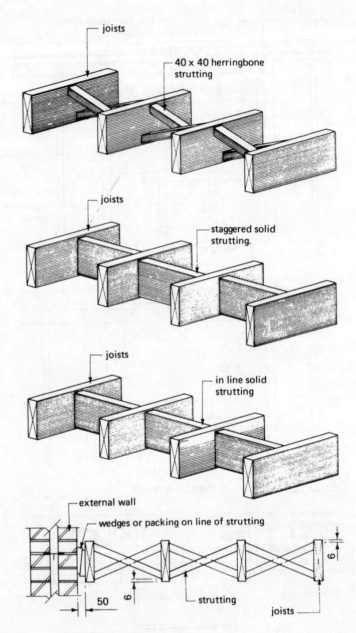

Fig. III.9   Strutting arrangements

238

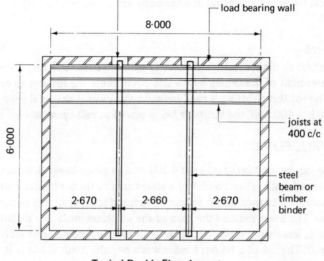

Typical Double Floor Layout

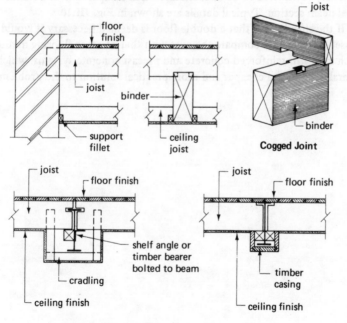

Cogged Joint

Typical Details Using Timber Binder or Steel Beam

Fig. III.10   Double floor details

included in modern designs, because they are considered to be superfluous. Typical trimming joints and arrangements are shown in Figs. III.7 and III.8

Strutting

Shrinkage in timber joists will cause twisting to occur and this will result in movement of the ceiling below and could cause the finishes to crack. To prevent this strutting is used between the joists if the total span exceeds 2 400 mm; the strutting being placed at mid-span (se Fig. III.9).

DOUBLE FLOORS

These can be used on spans over 4 500 mm to give a lower floor area free of internal walls. They consist of a steel beam or timber binder spanning the shortest distance which supports common joists spanning at right-angles. The beam reduces the span of the common joists to a distance which is less than the shortest span to allow an economic joist section to be used. The use of a timber binder was a popular method but it is generally considered to be uneconomic when compared with a standard steel beam section. Typical details are shown in Fig. III.10.

If the span is such that a double floor is deemed necessary it would be a useful exercise to compare the cost with that of other flooring methods, such as *in situ* reinforced concrete and precast concrete systems, which, overall, could be a cheaper and more practical solution to the problem.

24
Precast concrete floors

The function of any floor is to provide a level surface which is capable of supporting all the live and dead loads imposed. Reinforced concrete with its flexibility in design, good fire resistance and sound insulating properties is widely used for the construction of suspended floors for all types of buildings. The disadvantages of *in situ* concrete are:

1. Need for formwork.
2. Time taken for the concrete to cure before the formwork can be released for reuse and the floor available as a working area.
3. Very little is contributed by a large proportion of the concrete to the strength of the floor.

Floors composed of reinforced precast concrete units have been developed over the years to overcome some or all of the disadvantages of *in situ* reinforced concrete slab. To realise the full economy of any one particular precast flooring system the design of the floors should be within the span, width, loading and layout limitations of the units under consideration; coupled with the advantages of repetition.

CHOICE OF SYSTEM

Before any system of precast concrete flooring can be considered in detail the following factors must be taken into account:

1. Maximum span.
2. Nature of support.
3. Weight of units.

241

4. Thickness of units.
5. Thermal insulation properties.
6. Sound insulation properties.
7. Fire resistance of units.
8. Speed of construction.
9. Amount of temporary support required.

The systems available can be considered as either precast hollow floors or composite floors; further subdivision is possible by taking into account the amount of temporary support required during the construction period.

Precast hollow floors

Precast hollow floor units are available in a variety of sections such as box planks or beams; tee sections, I beam sections and channel sections (see Fig. III.11). The economies which can be reasonably expected over the *in situ* floor are:

1. 50% reduction in the volume of concrete.
2. 25% reduction in the weight of reinforcement.
3. 10% reduction in size of foundations.

The units are cast in precision moulds, around inflatable formers or foamed plastic cores. The units are laid side by side with the edge joints being grouted together; a structural topping is-not required but the upper surface of the units are usually screeded to provide the correct surface for the applied finishes (see Fig. III.11). Little or no propping is required during the construction period but usually some means of mechanical lifting is required to off load and position the units. Hollow units are normally the cheapest form of precast concrete suspended floor for simple straight spans with beam or wall supports up to a maximum span of 20.000 m. They are not considered suitable where heavy point loads are encountered unless a structural topping is used to spread the load over a suitable area.

The hollow beams or planks give a flat soffit which can be left in its natural state or be given a skim coat of plaster; the voids in the units can be used to house the services which are normally incorporated in the depth of the floor. The ribbed soffit of the channel and tee units can be masked by a suspended ceiling; again the voids created can be utilised to house the services. Special units are available with fixing inserts for suspended ceilings, service outlets and edges to openings.

Composite floors

These floors are a combination of precast units and *in situ* concrete. The precast units which are usually prestressed or reinforced with high yield steel bars are used to provide the strength of

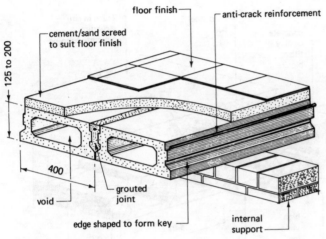

floor finish

anti-crack reinforcement

cement/sand screed
to suit floor finish

125 to 200

400

void

grouted
joint

edge shaped to form key

internal
support

spans up to 13·000

Typical hollow floor unit details

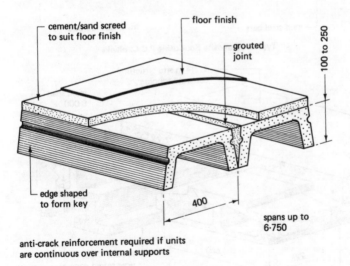

cement/sand screed
to suit floor finish

floor finish

grouted
joint

100 to 250

edge shaped
to form key

400

spans up to
6·750

anti-crack reinforcement required if units
are continuous over internal supports

Typical channel section floor unit details

Fig. III.11   Precast concrete hollow floors

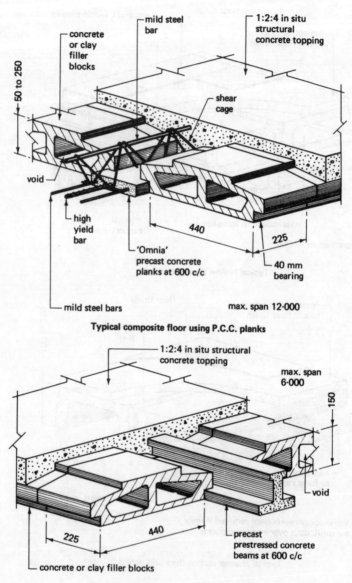

Typical composite floor using P.C.C. planks

Typical composite floor using P.C.C. beams

Fig. III.12   Composite floors

the floor with the smallest depth practicable and at the same time act as permanent formwork to the *in situ* topping which provides the compressive strength required. It is essential that an adequate bond is achieved between the two components – in most cases this is provided by the upper surface texture of the precast units; alternatively a mild steel fabric can be fixed over the units before the *in situ* topping is laid.

Composite floors generally take one of two forms:

1. Thin prestressed planks with a side key and covered with an *in situ* topping.
2. Reinforced or prestressed narrow beams which are placed at 600 mm centres and are bridged by concrete filler blocks, the whole combination being covered with *in situ* topping. Most of the beams used in this method have a shear reinforcing cage projecting from the precast beam section (see Fig. III.12).

In both forms temporary support should be given to the precast units by props at 1.800 to 2.400 m centres until the *in situ* topping has cured.

COMPARISON OF SYSTEMS

Precast hollow floors are generally cheaper than composite, *in situ* concrete is not required and therefore the need for mixing plant and storage of materials is eliminated. The units are self centering, therefore temporary support is not required, the construction period is considerably shorter and generally the overall weight is less.

Composite floors will act in the same manner as an *in situ* floor and can therefore be designed for more complex loadings. The formation of cantilevers is easier with this system and support beams can be designed within the depth of the floor giving a flat soffit. Services can be housed within the structural *in situ* topping, or within the voids of the filler blocks. Like the precast hollow floor, composite floors are generally cheaper than a comparable *in situ* floor within the limitations of the system employed.

25
Hollow block and waffle floors

Precast concrete suspended floors are generally considered to be for light to medium loadings spanning in one direction. Hollow block, or hollow pot floors as they are sometimes called, and waffle or honeycomb floors can be used as an alternative to the single spanning precast floor since they can be designed to carry heavier loadings. They are in fact ribbed floors consisting of closely spaced narrow and shallow beams giving an overall reduction in depth of the conventional reinforced concrete *in situ* beam and slab floor.

Hollow block floors

These are formed by laying over conventional floor soffit formwork a series of hollow lightweight clay blocks or pots in parallel rows with a space between these rows to form the ribs. The blocks act as permanent formwork giving a flat soffit suitable for plaster application and impart to the floor good thermal insulation and fire resistance. The ribs formed between the blocks can be reinforced to suit the loading conditions of the floor, thus providing flexibility of design (see Fig. III.13). The main advantages of this system are its light weight, which is generally less than comparable floors of concrete construction, and its relatively low cost.

Waffle or honeycomb floors

Used mainly as an alternative to an *in situ* flat slab or a beam and slab suspended floor since it requires less concrete,

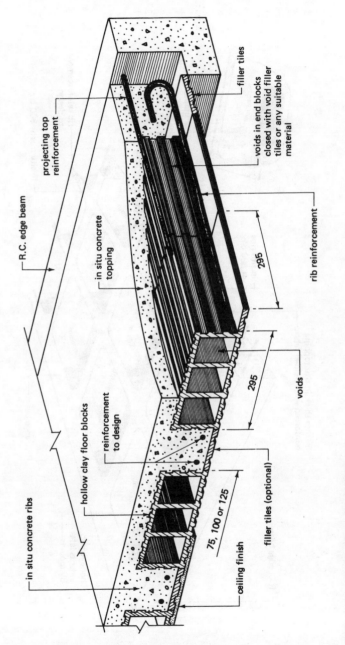

projecting top reinforcement

R.C. edge beam

in situ concrete topping

hollow clay floor blocks

reinforcement to design

in situ concrete ribs

filler tiles (optional)

ceiling finish

75, 100 or 125

voids

295

295

rib reinforcement

voids in end blocks closed with void filler tiles or any suitable material

filler tiles

Fig. III.13 Hollow block floors

247

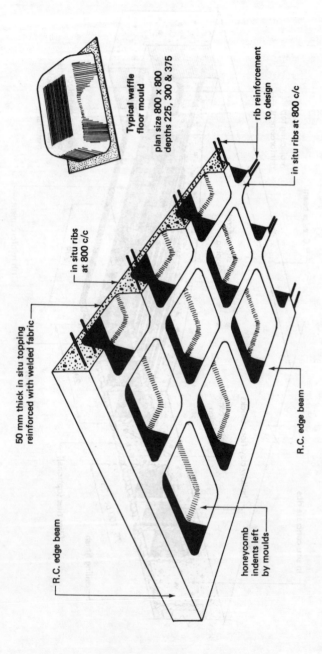

Typical waffle floor mould

plan size 800 x 800
depths 225, 300 & 375

rib reinforcement to design

in situ ribs at 800 c/c

in situ ribs at 800 c/c

50 mm thick in situ topping reinforced with welded fabric

R.C. edge beam

honeycomb indents left by moulds

R.C. edge beam

Fig. III.14   Waffle or honeycomb floors

less reinforcement and can be used to reduce the number of beams and columns required with the resultant savings on foundations. The honeycomb pattern on the underside can add to the visual aspect of the ceiling by casting attractive shadow patterns.

The floor is cast over lightweight moulds or pans made of glass fibre, polypropylene or steel forming a two-directional ribbed floor (see Fig. III.14). The moulds are very strong, lightweight and are capable of supporting all the normal loads encountered in building works. Support is reduced to the minimum since the moulds are arranged in parallel rows and span between the parallel lines of temporary supports.

The reinforcement in the ribs is laid in two directions to resist both longitudinal and transverse bending moments in the slab. Generally three mould depths are available but the overall depth can be increased by adding to the depth of the topping.

With all floors using an *in situ* topping it is possible to float the surface in preparation for the applied finishes, but this surface may suffer damage whilst being used by the following trades. It is therefore considered a better form of construction to allow for a floor screed to be applied to the *in situ* topping at a later stage in the contract prior to the fixing of the applied finish.

Part IV
Stairs

26
Timber stairs I

A stair is a means of providing access from one floor level to another. Modern stairs with their handrails are designed with the main emphasis on simplicity, trying to avoid the elaborate and costly features used in the past.

The total rise of a stair, that is, the distance from floor finish to floor finish in any one storey height, is fixed by the storey heights and floor finishes being used in the building design; therefore the stair designer only has the total going or total horizontal distance with which to vary his stair lay-out. It is good practice to keep door openings at least 450 mm away from the head or the foot of a stairway and to allow at least the stair width as circulation space at the head or foot of the stairway.

Stairs can be designed as one straight flight between floor levels which is the simplest and cheapest layout, alternatively they can be designed to turn corners by the introduction of quarter space (90°) or half space (180°) intermediate landings. Stairs which change direction of travel by using tapered steps or are based on geometrical curves in plan are beyond the scope of first year construction technology and are not considered in this text. Irrespective of the plan lay-out the principles of stair construction remain constant and are best illustrated by studying the construction of simple straight flight stairs; this is the principle followed in this volume.

Stair terminology
Stairwell: the space in which the stairs and landings are housed.

253

Stairs: the actual means of ascension or descension from one level to another.

Tread: the upper surface of a step on which the foot is placed.

Nosing: the exposed edge of a tread. usually projecting with a square. rounded or splayed edge.

Riser: the vertical member between two consecutive treads.

Step: riser plus tread.

Going: the horizontal distance between two consecutive risers or, as defined in Approved Document K, the distance measured on plan between two consecutive nosings.

Rise: the vertical height between two consecutive treads.

Flight: a series of steps without a landing.

Newel: post forming the junction of flights of stairs with landings or carrying the lower end of strings.

Strings: the members receiving the ends of steps which are generally housed to the string and secured by wedges, called wall or outer strings according to their position.

Handrail: protecting member usually parallel to the string and spanning between newels. This could be attached to a wall above and parallel to a wall string.

Baluster: the vertical infill member between a string and handrail.

Pitch line: a line connecting the nosings of all treads in any one flight.

BUILDING REGULATIONS PART K

These studies cover only those stairs intended for use in connection with dwellings.

Approved Document K defines two forms of stairs:

1. **Common stairway:** which is an internal or external stairway which forms part of a building and is intended for common use in connection with two or more dwellings.

2. **Private stairway:** which is an internal or external stairway of steps with straight nosings on plan which forms part of a building and is either within a dwelling or intended for use solely in connection with one dwelling.

Construction and design

It is essential to keep the dimensions of the treads and risers constant throughout any flight of steps to reduce the risk of accidents by changing the rhythm of movement up or down the stairway. The height of the individual step rise is calculated by dividing the total rise by the chosen number of risers. The individual step going is chosen to suit the floor area available so that it, together with the rise, meets the requirements of the Building Regulations (see Fig. IV.1). It is important to note that in any one flight there will be one more riser than treads since the last tread is in fact the landing.

Stairs are constructed by joining the steps into the spanning members or strings by using housing joints, glueing and wedging the steps into position to form a complete and rigid unit. Small angle blocks can be glued at the junction of tread and riser in a step to reduce the risk of slight movement giving rise to the annoyance of creaking. The flight can be given extra rigidity by using triangular brackets placed under the steps on the centre line of the flight. The use of a central beam or carriage piece with rough brackets as a support is only used on wide stairs over 1 200 mm, especially where they are intended for use as a common stairway (see Fig. IV.2.

Stairs can be designed to be either fixed to a wall with one outer string, fixed between walls or freestanding—the majority have one wall string and one outer string. The wall string is fixed directly to the wall along its entire length or is fixed to timber battens plugged to the wall, the top of the string being cut and hooked over the trimming member of the stairwell. The outer string is supported at both ends by a newel post which in the case of the bottom newel rests on the floor; and, in the case of the upper newel, it is notched over and fixed to the stairwell trimming member. If the upper newel is extended to the ground floor to give extra support, it is called a storey newel post. The newel posts also serve as the termination point for handrails which span between them and is then infilled with balusters, balustrade or a solid panel to complete the protection to the sides of the stairway (see Fig. IV.3).

If the headroom distance is critical it is possible to construct a bulk-head arrangement over the stairs as shown in Fig. IV.4; this may give the increase in headroom required to comply with the Building Regulations. The raised floor in the room over the stairs can be used as a shelf or form the floor of a hanging cupboard.

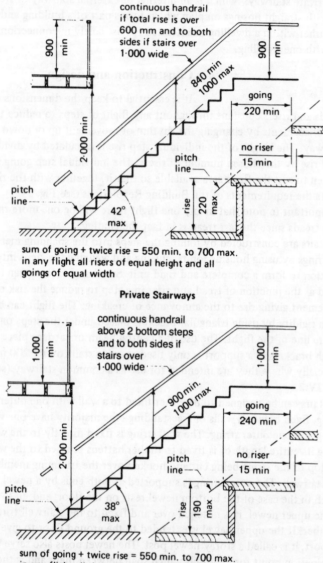

continuous handrail
if total rise is over
600 mm and to both
sides if stairs over
1·000 wide

900 min

900 min

840 min
1000 max

2·000 min

pitch line

pitch line

going
220 min

no riser
15 min

pitch line

rise 220 max

42° max

sum of going + twice rise = 550 min. to 700 max.
in any flight all risers of equal height and all
goings of equal width

Private Stairways

continuous handrail
above 2 bottom steps
and to both sides if
stairs over 1·000 wide

1·000 min

1·000 min

900 min
1000 max

2·000 min

pitch line

pitch line

going
240 min

no riser
15 min

pitch line

rise 190 max

38° max

sum of going + twice rise = 550 min. to 700 max.
in any flight all risers of equal height and all
goings of equal width.

Common Stairways

Fig. IV.1   Timber stairs and Approved Document K

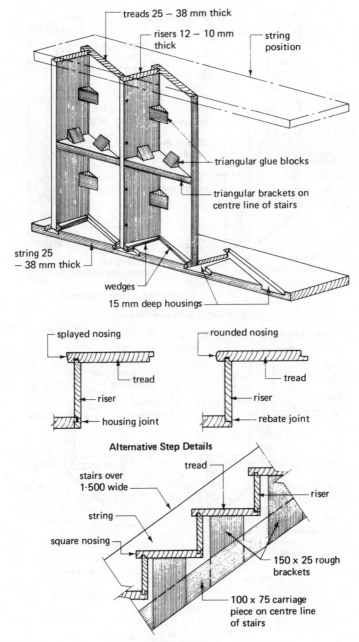

treads 25 – 38 mm thick

risers 12 – 10 mm thick

string position

triangular glue blocks

triangular brackets on centre line of stairs

string 25 – 38 mm thick

wedges

15 mm deep housings

splayed nosing

tread

riser

housing joint

rounded nosing

tread

riser

rebate joint

Alternative Step Details

tread

stairs over 1·500 wide

riser

string

square nosing

150 x 25 rough brackets

100 x 75 carriage piece on centre line of stairs

Fig. IV.2   Stair construction details

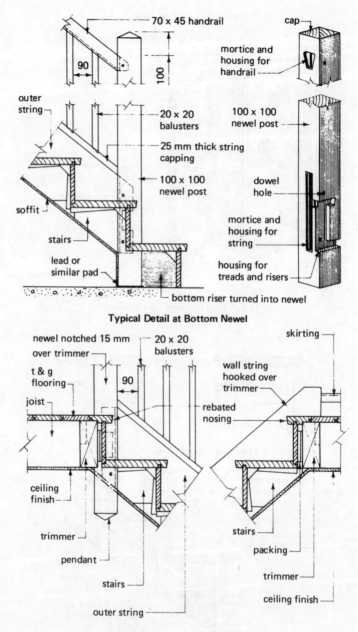

70 x 45 handrail

cap

90

100

mortice and
housing for
handrail

outer
string

20 x 20
balusters

25 mm thick string
capping

100 x 100
newel post

100 x 100
newel post

soffit

dowel
hole

stairs

mortice and
housing for
string

lead or
similar pad

housing for
treads and risers

bottom riser turned into newel

Typical Detail at Bottom Newel

newel notched 15 mm
over trimmer

20 x 20
balusters

skirting

t & g
flooring

90

wall string
hooked over
trimmer

joist

rebated
nosing

ceiling
finish

trimmer

stairs

pendant

packing

stairs

trimmer

outer string

ceiling finish

Typical Details at Landing

Fig. IV.3 Stair support and fixing details

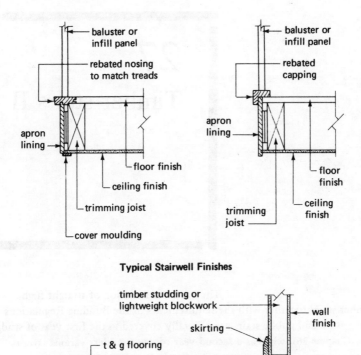

Typical Stairwell Finishes

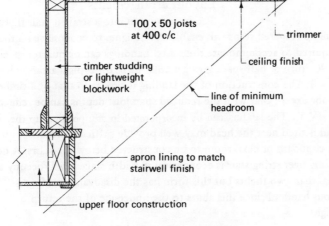

Typical Bulkhead Over Stairs

Fig. IV.4 Stairwell finishes and bulkhead details

27
Timber stairs II

The basic construction of straight flight timber stairs together with the requirements of the Building Regulations for private and public stairs are generally covered in the first year of study (see Chapter 26), whereas a second year course considers various layout arrangements and open tread or riserless stairs.

Layout arrangements

The introduction into a straight stair flight of landings or tapered steps will enable the designer to economise on the space required to accommodate the stairs. Landings can be quarter space giving a 90° turn or half space giving a 180° turn; for typical arrangements see Fig. V.1. The construction of the landing is similar to that of a timber upper floor except that with the reduced span joist depths can be reduced (see Fig. IV.5). The landing can be incorporated in any position up the flight and if sited near the head may well provide sufficient headroom to enable a cupboard or cloakroom to be constructed below the stairs. A dog leg or string over string stair is economical in width since it will occupy a width less than two flights but this form has the disadvantage of a discontinuous handrail since this abuts to the underside of the return or upper flight.

Tapered steps

Prior to the introduction of the Building Regulations tapered steps or winders were frequently used by designers to use space economically since three treads occupied the area required for the conventional quarter space landing which is counted as one tread. These steps had the following disadvantages:

1. Hazard to the aged and very young because of the very small tread length at or near the newel post.
2. Difficult to carpet, requiring many folds or wasteful cutting.
3. Difficult to negotiate with furniture due to a rapid rise on the turn.
4. Have little or no aesthetic appeal.
5. Are expensive to construct.

With the introduction of the Building Regulations special attention has been given to the inclusion of tapered steps in Approved Document K which makes the use of tapered steps less of an economic proposition and more difficult to design (see Fig. IV.6).

Open tread stairs

These are a contemporary form of stairs used in homes, shops and offices based on the simple form of access stair which has been used for many years in industrial premises. The concept is simplicity with the elimination of elaborate nosings, cappings and risers. The open tread or riserless stair must fully comply with Part K of the Building Regulations and in particular to Approved Document K which recommends a minimum tread overlap of 15 mm.

Three basic types of open tread stairs can be produced:

Closed string: which would be terminated at the floor and landing levels and fixed as for traditional stairs. The treads are tightly housed into the strings which are tied together with long steel tie bars under the first, last and every fourth tread. The nuts and washers can be housed into the strings and covered with timber inserts (see Fig. IV.7).

Cut strings or carriages: these are used to support cantilever treads and can be worked from the solid or of laminated construction. The upper end of the carriage can be housed into the stairwell trimming member with possible additional support from metal brackets. The foot of the carriage is housed in a purpose made metal shoe or fixed with metal angle brackets (see Fig. IV.8).

Mono-carriage: sometimes called a spine beam, employs a single central carriage with double cantilever treads. The carriage, which is by necessity large, is of laminated construction and very often of a tapered section to reduce the apparent bulky appearance. The foot of the carriage is secured with a purpose made metal shoe in conjunction with timber connectors (see Fig. IV.9).

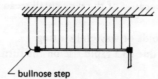

bullnose step

Straight flight

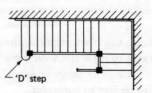

'D' step

Quarter turn

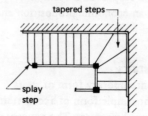

tapered steps

splay step

Quarter turn

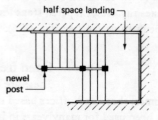

half space landing

newel post

Dog-leg or string over string

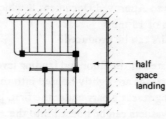

half space landing

Open well or open newel

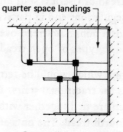

quarter space landings

Open well or open newel

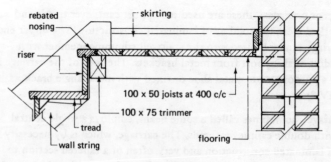

rebated nosing

skirting

riser

100 x 50 joists at 400 c/c

100 x 75 trimmer

tread

flooring

wall string

Landing details

Fig. IV.5 Typical stair layouts

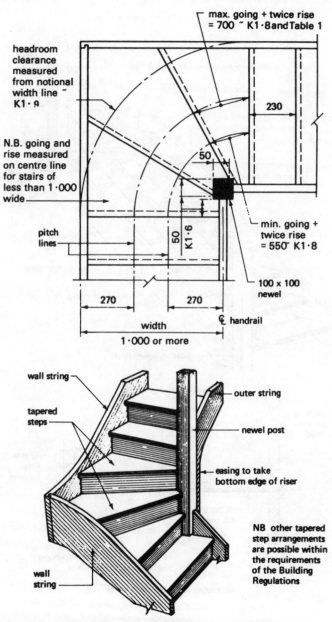

max. going + twice rise
= 700 ~ K1·8 and Table 1

headroom
clearance
measured
from notional
width line ~
K1·9

230

N.B. going and
rise measured
on centre line
for stairs of
less than 1·000
wide

50

min. going +
twice rise
= 550 ~ K1·8

pitch
lines

50

K1·6

100 × 100
newel

270   270

℄ handrail

width
1·000 or more

wall string

outer string

tapered
steps

newel post

easing to take
bottom edge of riser

wall
string

NB other tapered
step arrangements
are possible within
the requirements
of the Building
Regulations

Fig. IV.6   Tapered steps for private stairways

263

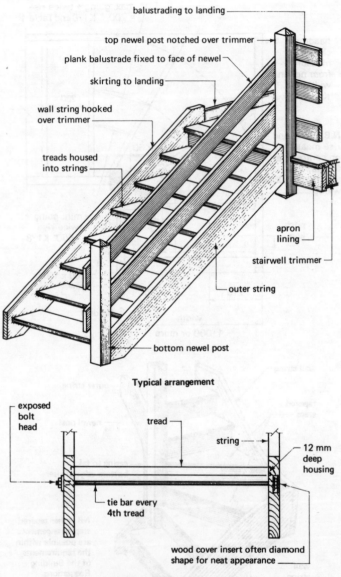

balustrading to landing

top newel post notched over trimmer

plank balustrade fixed to face of newel

skirting to landing

wall string hooked over trimmer

treads housed into strings

apron lining

stairwell trimmer

outer string

bottom newel post

Typical arrangement

exposed bolt head

tread

string

12 mm deep housing

tie bar every 4th tread

wood cover insert often diamond shape for neat appearance

Alternative tie bar arrangements

Fig. IV.7 Closed string open tread stairs

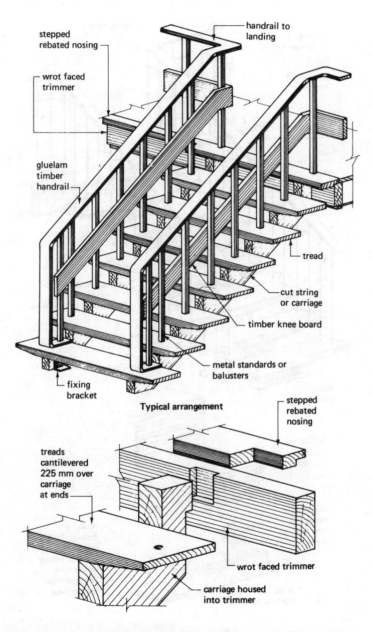

Typical arrangement

Fig. IV.8 Cut string open tread stairs

265

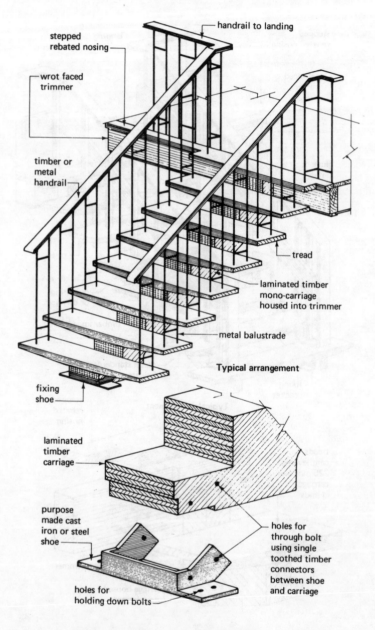

handrail to landing

stepped
rebated nosing

wrot faced
trimmer

timber or
metal
handrail

tread

laminated timber
mono-carriage
housed into trimmer

metal balustrade

Typical arrangement

fixing
shoe

laminated
timber
carriage

purpose
made cast
iron or steel
shoe

holes for
through bolt
using single
toothed timber
connectors
between shoe
and carriage

holes for
holding down bolts

Fig. IV.9 Mono-carriage open tread stairs

Treads

These must be of adequate thickness since there are no risers to give extra support; usual thicknesses are 38 and 50 mm. To give a lighter appearance it is possible to taper the underside of the treads at their cantilever ends for a distance of 225–250 mm. This distance is based on the fact that the average person will ascend or descend a stairway at a distance of about 250 mm in from the handrail.

Balustrading

Together with the handrail balustrading provides both the visual and practical safety barrier to the side of the stair. Children present special design problems since they can and will explore any gap big enough to crawl through. BS 5395 for wood stairs recommends that the infill under handrails should have no openings which would permit the passage of a sphere 90 mm in diameter. Many variations of balustrading are possible ranging from simple newels with planks to elaborate metalwork of open design (see Figs. IV.7, IV.8, and IV.9).

28
Simple reinforced concrete stairs

The functions of any stairway are:

1. To provide for the movement of people from one floor level to another.
2. To provide a suitable means of escape in case of fire.

A timber stairway will maintain its strength for a reasonable period during an outbreak of fire but will help to spread the fire thus increasing the hazards which could be encountered along an escape route. It is for this reason that the use of timber stairs are limited by Building Regulations B2 and B3 to certain domestic building types and stairs within shops which are not in a protective shaft and are not therefore part of the planned fire escape route.

Stairs, other than the exceptions given above, must therefore be constructed of non-combustible materials, but combustible materials are allowed to be used as finishes to the upper surface of the stairway or landing. Reinforced concrete stairs are non-combustible, strong and hard wearing. They may be constructed *in situ* or precast in sections ready for immediate installation and use when delivered to site. The general use of cranes on building sites has meant that many of the large flight arrangements, which in the past would have been cast *in situ*, can now be precast under factory controlled conditions.

Many variations of plan layout and spanning direction are possible but this study will be confined to the simple straight flight spanning from floor to floor or floor to intermediate landing. The designer will treat the stair as being an inclined slab spanning simply between the supports, the steps being treated as triangular loadings evenly distributed over the length.

Where intermediate landings are included in the design the basic plan is similar to the open well or newel timber stair. Difficulty is sometimes experienced with the intersection of the upper and lower flight soffits with the landing. One method of overcoming this problem is to have, in plan, the top and bottom steps out of line so that the soffit intersections form a straight line on the underside of the landing (see Fig. IV.10). The calculations to determine the rise, going and number of steps is the same as those used for timber stairs; it should be noted that the maximum number of risers in each flight must not exceed sixteen. To achieve a greater tread length without increasing the actual horizontal going it is common to use a splayed riser face giving a 25 mm increase to the tread length.

The concrete specification is usually 1 : 2 : 4/20 mm aggregate with a cover of concrete over the reinforcement of 15 mm minimum or the bar diameter, whichever is the greater. This cover will give a one-hour fire resistance which is the minimum period implied by the Building Regulations. The thickness of concrete required is dependent on the loading and span but is not generally less than 100 mm or more than 150 mm measured across the waist which is the distance from the soffit to the intersection of tread and riser (see Fig. IV.10).

Mild steel or high yield steel bars can be used to reinforce concrete stairs, the bars being lapped to starter bars at the ground floor and taken into the landing or floor support slab. The number, diameter and spacing of the main and distribution reinforcement must always be calculated for each stairway.

Handrails and balustrading must be constructed of a non-combustible material, continuous and to both sides if the width of the stairs exceeds 1.000 m. The overall height of the handrail up the stairs should be between 900 and 1 000 mm measured vertically and have a height above the floor of 1.100 m minimum. The capping can be of a combustible material such as plastic provided that it is fixed to or over a non-combustible core. Methods of securing balustrades and typical handrail details are shown in Fig. IV.11.

A wide variety of finishes can be applied to the tread surface of the stairs. If the appearance is not of paramount importance, such as in a warehouse, a natural finish could be used but it would be advisable to trowel into the surface some carborundum dust to provide a hard-wearing non-slip surface. Alternatively, rubber or carborundum insert strips could be fixed or cast into the leading edges of the treads. Finishes such as PVC tiles, rubber tiles, and carpet mats are applied and fixed in the same manner as for floors. The soffits can be left as struck from the formwork and decorated or finished with a coat of spray plaster or a skim coat of finishing plaster.

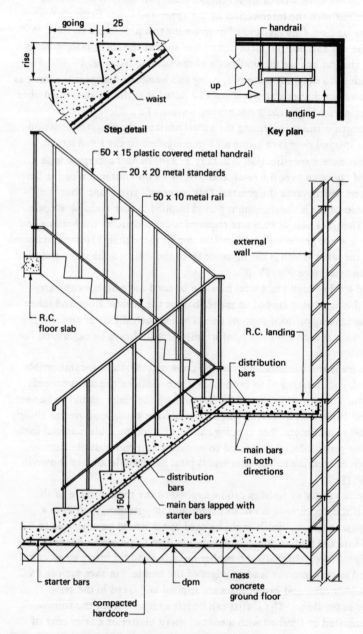

going

25

rise

waist

Step detail

handrail

up

landing

Key plan

50 x 15 plastic covered metal handrail

20 x 20 metal standards

50 x 10 metal rail

external wall

R.C. floor slab

R.C. landing

distribution bars

main bars in both directions

distribution bars

main bars lapped with starter bars

150

starter bars

dpm

mass concrete ground floor

compacted hardcore

Fig. IV.10 Simple R.C. *in situ* stairs

270

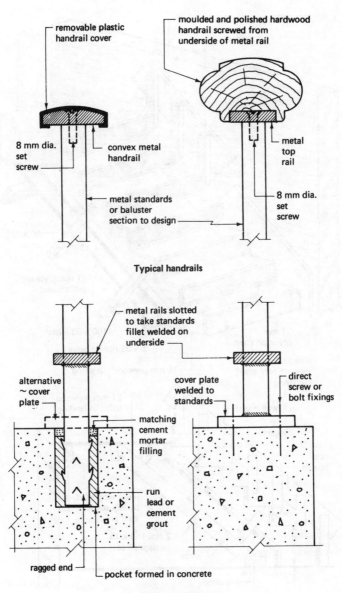

removable plastic
handrail cover

moulded and polished hardwood
handrail screwed from
underside of metal rail

8 mm dia.
set
screw

convex metal
handrail

metal
top
rail

metal standards
or baluster
section to design

8 mm dia.
set
screw

Typical handrails

metal rails slotted
to take standards
fillet welded on
underside

alternative
~ cover
plate

cover plate
welded to
standards

direct
screw or
bolt fixings

matching
cement
mortar
filling

run
lead or
cement
grout

ragged end

pocket formed in concrete

Typical fixing methods

Fig. IV.11 Handrails and balustrades

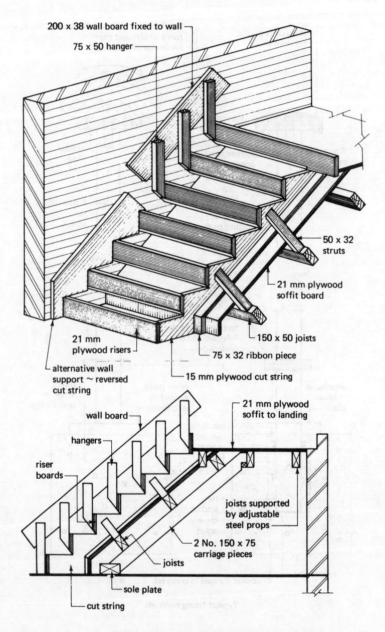

Fig. IV.12 **Typical formwork to R.C. *in situ* stairs**

Formwork

The basic requirements are the same as for formwork to a framed structure. The stair profile is built off an adequately supported soffit of sheet material by using a cut string. Riser boards are used to form the leading face of the steps; these should have a splayed bottom edge to enable complete trowelling of the tread surfaces and to ensure that air is not trapped under the bottom edge of the riser board thus causing voids. If the stair abuts a vertical surface two methods can be considered to provide the abutment support for the riser boards; a reversed cut string or a wall board with hangers (see Fig. IV.12). Wide stairs can have a reverse cut string as a central support to the riser boards to keep the thickness of these within an acceptable coat limit.

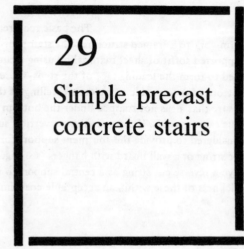

29
Simple precast concrete stairs

Precast concrete stairs can be designed and constructed to satisfy a number of different requirements. They can be a simple inclined slab, a cranked slab, an open riser stair or constructed from a series of precast steps built into, and if required cantilevered from, a structural wall.

The design considerations for the simple straight flight are the same as those for *in situ* stairs of comparable span, width and loading conditions. The fixing and support, however, require a different approach. Bearings for the ends of the flights must be provided at the floor or landing levels in the form of a haunch, rebate or bracket and continuity of reinforcement can be achieved by leaving projecting bars and slots in the floor into which they can be grouted (see Fig. IV.13).

Ideally the delivery of precast stairs should be arranged so that they can be lifted, positioned and fixed direct from the delivery vehicle, thus avoiding double handling. Precast components are usually designed for two conditions:

1. Lifting and transporting.
2. Final fixed condition.

It is essential that the flights are lifted from the correct lifting points, which may be in the form of loops or hooks projecting from or recessed into the concrete member, if damage by introducing unacceptable stresses during lifting are to be avoided.

Balustrade and handrail requirements together with the various methods of fixing are as described for *in situ* reinforced concrete stairs. Any tread

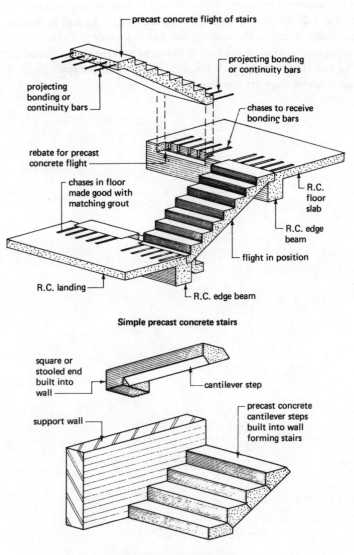

- precast concrete flight of stairs

projecting bonding
or continuity bars

projecting
bonding or
continuity bars

chases to receive
bonding bars

rebate for precast
concrete flight

chases in floor
made good with
matching grout

R.C.
floor
slab

R.C. edge
beam

flight in position

R.C. landing

R.C. edge beam

Simple precast concrete stairs

square or
stooled end
built into
wall

cantilever step

support wall

precast concrete
cantilever steps
built into wall
forming stairs

Simple precast concrete steps

Fig. IV.13   Precast concrete stairs and steps

finish which is acceptable for an *in situ* stair will also be suitable for the precast alternative.

The use of precast concrete steps to form a stairway are limited to situations such as short flights between changes in floor level and external stairs to basements and areas. They rely on the load bearing wall for support and if cantilevered on the downward load of the wall to provide the necessary reaction. The support wall has to provide this necessary load and strength, and at the same time it has to be bonded or cut around the stooled end(s) of the steps. It is for these reasons that the application of precast concrete steps are restricted.

30
Concrete stairs

Any form of stairs is primarily a means of providing circulation and communication between the various levels within a building. Apart from this primary function stairs may also be classified as a means of escape in case of fire; if this is the case the designer is severely limited by necessary regulations as to choice of materials, position and sizing of the complete stairway. Stairs which do not fulfil this means of escape function are usually called accommodation stairs and as such are not restricted by the limitations given above for escape stairs.

Escape stairs have been covered in a previous section of this volume and it is only necessary to reiterate the main points:

1. Constructed from non-combustible materials.
2. Stairway protected by a fire-resisting enclosure.
3. Separated from the main floor area by a set or sets of self-closing fire-resisting doors.
4. Limitations as to riser heights, tread lengths and handrail requirement usually based upon use of building.
5. In common with all forms of stairs, all riser heights must be equal throughout the rise of the stairs.

It should be appreciated that points 2 and 3 listed above can prove to be inconvenient to persons using the stairway for general circulation within the building, such as having to pass through self-closing doors, but in the context of providing a safe escape route this is unavoidable.

Generally concrete escape stairs are designed in straight flights with not more than 16 risers per flight providing such flights are not constructed in line. A turn of at least 30° is required after any flight which has more than thirty-six risers. In special circumstances such as low usage (not more than 50 persons), minimum overall diameter of 1.500 m and a total rise not exceeding 3.000 m spiral stairs may be acceptable as a suitable means of escape in case of fire.

The construction of simple reinforced concrete stairs together with a suitable formwork arrangement is usually covered in the second year of a course in construction technology (see Chapter 28) and it is a useful exercise, at this stage, for the student to refresh his memory on the following basic requirements:

1. Concrete mix usually specified as 1:2:4/20 mm aggregate.
2. Minimum cover of concrete over reinforcement 15 mm or bar diameter whichever is the greater to give a 1-hour fire resistance.
3. Waist thickness usually between 100 and 250 mm depending on stair type.
4. Mild steel or high yield steel bars can be used as reinforcement.
5. Continuous handrails of non-combustible materials at a height of 900 mm above the pitch line are required to all stairs and to both sides if the stair width exceeds 1.000 m.

The main area of study for an advanced level course is concerned with the different stair types or arrangements which can range from a single straight flight to an open spiral stair.

Single straight flight stair: this form of stair, although simple in design and construction, is not popular because of the plan space it occupies. In this stair arrangement the flight behaves as a simple supported slab spanning from landing to landing. The effective span or total horizontal going is usually taken as being from landing edge to edge by providing a downstand edge beam to each landing. If these edge beams are not provided the effective span would be taken as overall of the landings, resulting in a considerably increased bending moment and hence more reinforcement. Typical details are shown in Fig. IV.14.

Inclined slab stair with half space landings: these stairs have the usual plan format for reinforced concrete stairs giving a more compact plan layout and better circulation than the single straight flight stairs. The half space or 180° turn landing is usually introduced at the mid-point of the rise giving equal flight spans, thus reducing the effective span and hence the bending moment considerably. In most designs the landings span crosswise on to a load bearing wall or beam and the flights span from landing to landing. The point of intersection of the soffits to the flights with the landing soffits can be detailed in one of two ways:

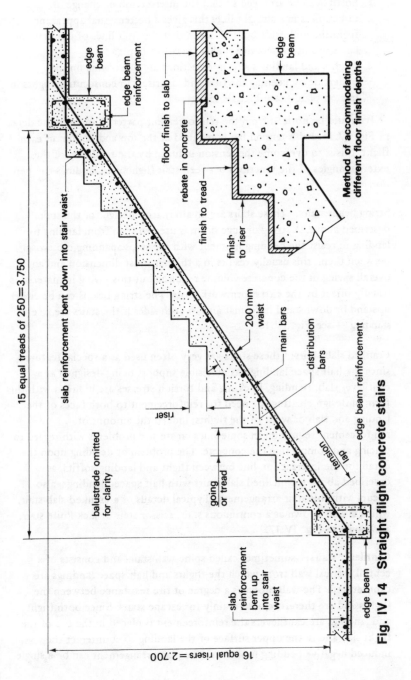

Fig. IV.14 Straight flight concrete stairs

edge beam

edge beam reinforcement

slab reinforcement bent down into stair waist

15 equal treads of 250 = 3.750

balustrade omitted for clarity

200 mm waist

main bars

distribution bars

riser

going or tread

tension

lap

edge beam reinforcement

slab reinforcement bent up into stair waist

edge beam

16 equal risers = 2.700

floor finish to slab

rebate in concrete

finish to tread

finish to riser

edge beam

Method of accommodating different finish depths

1. Soffits can be arranged so that the intersection or change of direction is in a straight line; this gives a better visual appearance from the underside but will mean that the riser lines of the first and last steps in consecutive flights are offset in plan.
2. Flights and landing soffit intersections are out of line on the underside by keeping the first and last risers in consecutive flights in line on plan — see Fig. IV.15.

It should be noticed from the reinforcement pattern shown in the detail in Fig. IV.15 that a tension lap is required at the top and bottom of each flight, this is to overcome the tension induced by the tendency of the external angles of the junctions between stair flights and landings to open out.

String beam stairs: these stairs are an alternative design for the stairs described above. A string or edge beam is used to span from landing to landing to resist the bending moment with the steps spanning crosswise between them; this usually results in a thinner waist dimension and an overall saving in the concrete volume required but this saving in material is usually offset by the extra formwork costs. The string beams can be either upstand or downstand in format and to both sides if the stairs are free standing — see Fig. IV.16.

Cranked slab stairs: these stairs are very often used as a special feature since the half space landing has no visible support being designed as a cantilever slab. Bending, buckling and torsion stresses are induced with this form of design creating the need for reinforcement to both faces of the landing and slab or waist of the flights; indeed the amount of reinforcement required can sometimes create site problems with regard to placing and compacting the concrete. The problem of deciding upon the detail for the intersection line between flight and landing soffits, as described above for inclined slab stairs with half spaced landings, also occurs with this stair arrangement. Typical details of a cranked slab stair, which is also known as a continuous stair, scissor stair or jack knife stair, are shown in Fig. IV.17.

Cantilever stairs: sometimes called spine wall stairs and consists of a central vertical wall from which the flights and half space landings are cantilevered. The wall provides a degree of fire resistance between the flights and are therefore used mainly for escape stairs. Since both flights and landings are cantilevers the reinforcement is placed in the top of the flight slab and in the upper surface of the landing to counteract the induced negative bending moments. The plan arrangement can be a single

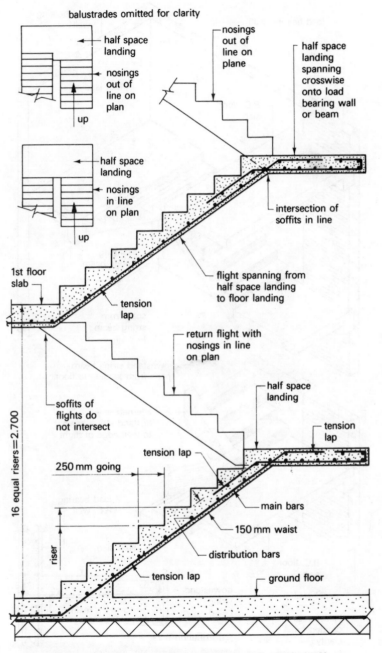

balustrades omitted for clarity

half space landing

nosings out of line on plan

up

half space landing

nosings in line on plan

up

nosings out of line on plane

half space landing spanning crosswise onto load bearing wall or beam

intersection of soffits in line

1st floor slab

tension lap

flight spanning from half space landing to floor landing

return flight with nosings in line on plan

half space landing

tension lap

soffits of flights do not intersect

16 equal risers = 2.700

250 mm going

tension lap

riser

main bars

150 mm waist

distribution bars

tension lap

ground floor

tension lap

Fig. IV.15   Inclined slab concrete stair with half space landings

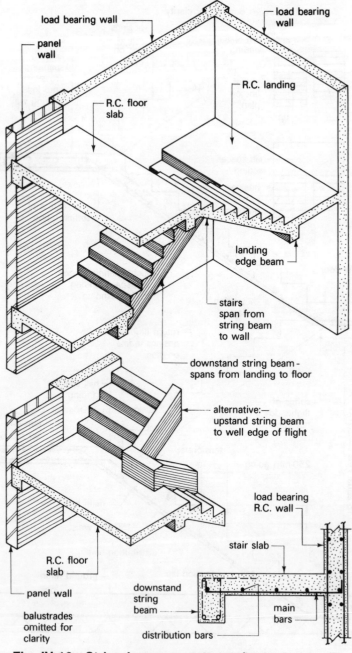

load bearing wall

panel wall

R.C. floor slab

R.C. landing

load bearing wall

landing edge beam

stairs span from string beam to wall

downstand string beam - spans from landing to floor

alternative:- upstand string beam to well edge of flight

R.C. floor slab

panel wall

balustrades omitted for clarity

load bearing R.C. wall

stair slab

downstand string beam

main bars

distribution bars

Fig. IV.16 String beam concrete stairs

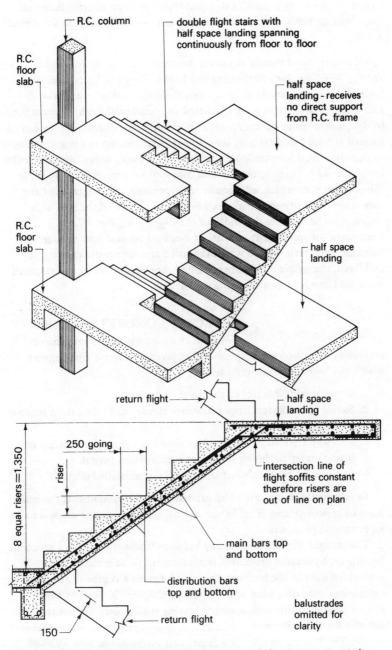

R.C. column

double flight stairs with half space landing spanning continuously from floor to floor

R.C. floor slab

half space landing - receives no direct support from R.C. frame

R.C. floor slab

half space landing

return flight

half space landing

250 going

riser

8 equal risers = 1.350

intersection line of flight soffits constant therefore risers are out of line on plan

main bars top and bottom

distribution bars top and bottom

150

return flight

balustrades omitted for clarity

Fig. IV.17 Cranked or continuous slab concrete stairs

straight flight or, as is usual, two equal flights with an intermediate half space landing between consecutive stair flights — see Fig. IV.18 for typical details.

Spiral stairs: used mainly as accommodation stairs in the foyers of prestige buildings such as theatres and banks. They can be expensive to construct being normally at least seven times the cost of conventional stairs. The plan shape is generally based on a circle although it is possible to design an open spiral stair with an elliptical core. The spiral stair can be formed around a central large diameter circular column in a similar manner to that described for cantilevered stair or, as is usual, design with a circular open stair well. Torsion, tension and compressive stresses are induced in this form of stair which will require reinforcement to both faces of the slab in the form of radial main bars bent to the curve of the slab with distribution bars across the width of the flight — see Fig. IV.19. Formwork for spiral stairs consists of a central vertical core or barrel to form the open stair well to which the soffit and riser boards are set out and fixed, the whole arrangement being propped and strutted as required from the floor level in a conventional manner.

PRECAST CONCRETE STAIRS

Most of the concrete stair arrangements previously described can be produced as precast concrete components which can have the following advantages:

1. Better quality control of the finished product.
2. Saving in site space, since formwork storage and fabrication space is no longer necessary.
3. Stairway enclosing shaft can be utilised as a space for hoisting or lifting materials during the major construction period.
4. Can usually be positioned and fixed by semi-skilled labour.

In common with the use of all precast concrete components the stairs must be repetitive and in sufficient quantity to justify their use and to be an economic proposition.

The straight flight stair spanning between landings can have a simple bearing or, by leaving projecting reinforcement to be grouted into preformed slots in the landings, they can be given a degree of structural continuity; this latter form is illustrated in Chapter 29. Straight flight precast concrete stairs with a simple bearing require only bottom reinforcement to the slab and extra reinforcement to strengthen the bearing rebate or nib. The bearing location is a rebate cast in the *in situ* floor slab or landing leaving a tolerance gap of 8 to 12 mm which is filled with a com-

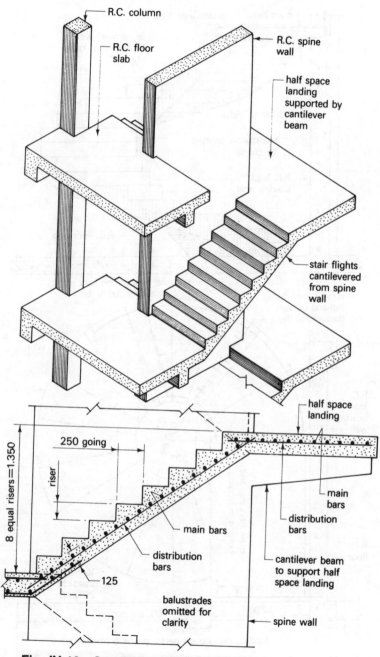

Fig. IV.18 Cantilever concrete stairs

R.C. column

R.C. floor slab

R.C. spine wall

half space landing supported by cantilever beam

stair flights cantilevered from spine wall

half space landing

main bars

distribution bars

cantilever beam to support half space landing

spine wall

250 going

riser

main bars

distribution bars

125

8 equal risers =1.350

balustrades omitted for clarity

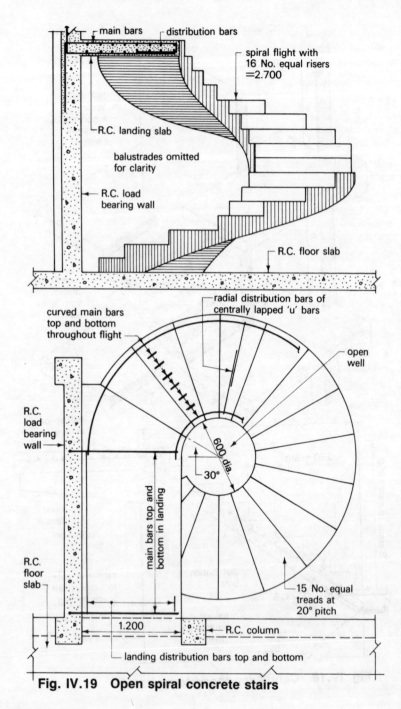

main bars   distribution bars

spiral flight with
16 No. equal risers
=2.700

R.C. landing slab

balustrades omitted
for clarity

R.C. load
bearing wall

R.C. floor slab

radial distribution bars of
centrally lapped 'u' bars

curved main bars
top and bottom
throughout flight

open
well

R.C.
load
bearing
wall

600 dia.

30°

R.C.
floor
slab

main bars top and
bottom in landing

1.200

R.C. column

15 No. equal
treads at
20° pitch

landing distribution bars top and bottom

Fig. IV.19 Open spiral concrete stairs

286

pressible material to form a flexible joint. The decision as to whether the stair and landing soffits will be in line or, alternatively, the first and last risers kept in line on plan remains, and since the bearing rebates are invariably cast in a straight line to receive both upper and lower stair flights the intersection design is detailed with the precast units — a typical example is shown in Fig. IV.20.

Cranked slab precast concrete stairs are usually formed as an open well stair. The bearing for the precast landings to the *in situ* floor or to the structural frame is usually in the form of a simple bearing as described above for straight flight precast concrete stairs. The infill between the two adjacent flights, in an open well plan arrangement at floor and intermediate landing levels, can be of *in situ* concrete with structural continuity provided by leaving reinforcement projecting from the inside edge of the landings — see Fig. IV.21. it must be remembered that when precast concrete stair flights are hoisted into position different stresses may be induced from those which will be encountered in the fixed position. To overcome this problem the designer can either reinforce the units for both conditions or as is more usual provide definite lifting points in the form of projecting lugs or by utilising any holes cast in to receive the balustrading.

Precast open riser stairs are a form of stair which can be both economic and attractive consisting of a central spine beam in the form of a cut string supporting double cantilever treads of timber or precast concrete. The foot of the lowest spine beam is located and grouted into a preformed pocket cast in the floor whereas the support at landing and floor levels is a simple bearing located in a housing cast into the slab edge — see Fig. IV.22. Anchor bolts or cement in sockets are cast into the spine beam to provide the fixing for the cantilever treads. The bolt heads are recessed below the upper tread surface and grouted over with a matching cement mortar for precast concrete treads and concealed with matching timber pellets when hardwood treads are used. The supports for the balustrading and handrail are located in the holes formed at the ends of the treads and secured with a nut and washer on the underside of the tread. It will have been noticed that the balustrade and handrail details have been omitted from all the stair details so far considered; this has been done for reasons of clarity and this aspect of stairwork is usually covered during the introductory work in the second year of a typical four year course of study in construction technology — see Chapter 29.

Spiral stairs in precast concrete work are based upon the stone stairs found in many historic buildings such as Norman castles and cathedrals, consisting essentially of steps which have a 'keyhole' plan shape rotating round a central core. Precast concrete spiral stairs are usually open riser

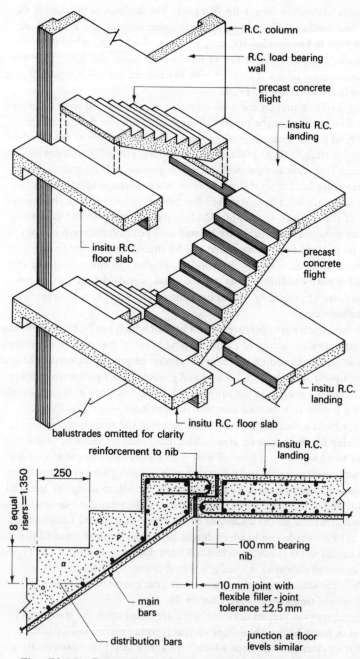

R.C. column

R.C. load bearing wall

precast concrete flight

insitu R.C. landing

insitu R.C. floor slab

precast concrete flight

insitu R.C. landing

insitu R.C. floor slab

balustrades omitted for clarity

insitu R.C. landing

reinforcement to nib

250

8 equal risers = 1.350

100 mm bearing nib

10 mm joint with flexible filler - joint tolerance ±2.5 mm

main bars

distribution bars

junction at floor levels similar

Fig. IV.20 Precast concrete straight flight stairs

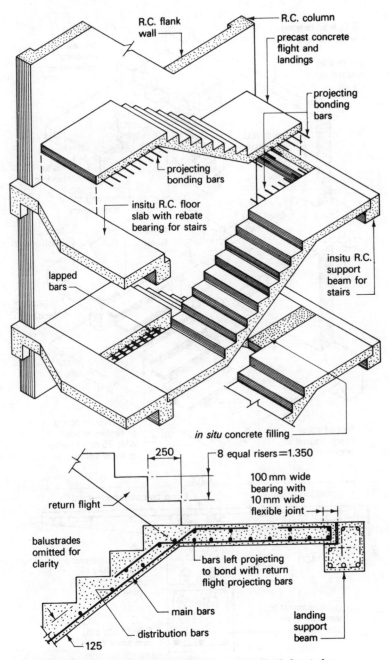

R.C. flank wall

R.C. column

precast concrete flight and landings

projecting bonding bars

projecting bonding bars

insitu R.C. floor slab with rebate bearing for stairs

insitu R.C. support beam for stairs

lapped bars

in situ concrete filling

250

8 equal risers = 1.350

100 mm wide bearing with 10 mm wide flexible joint

return flight

balustrades omitted for clarity

bars left projecting to bond with return flight projecting bars

main bars

distribution bars

landing support beam

125

Fig. IV.21   Precast concrete cranked slab stairs

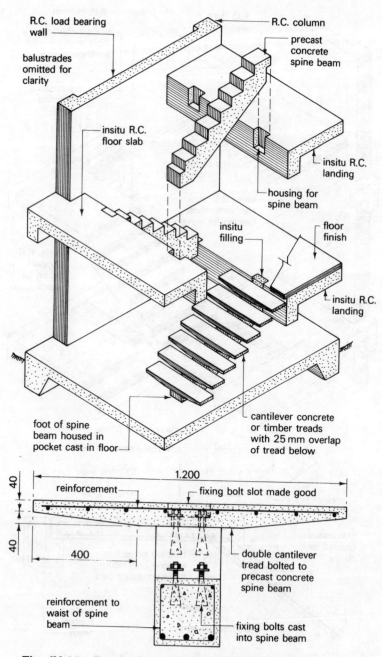

R.C. load bearing wall

balustrades omitted for clarity

R.C. column

precast concrete spine beam

insitu R.C. floor slab

insitu R.C. landing

housing for spine beam

insitu filling

floor finish

insitu R.C. landing

foot of spine beam housed in pocket cast in floor

cantilever concrete or timber treads with 25 mm overlap of tread below

1.200

40

reinforcement

fixing bolt slot made good

40

400

double cantilever tread bolted to precast concrete spine beam

reinforcement to waist of spine beam

fixing bolts cast into spine beam

Fig. IV.22 Precast concrete open riser stairs

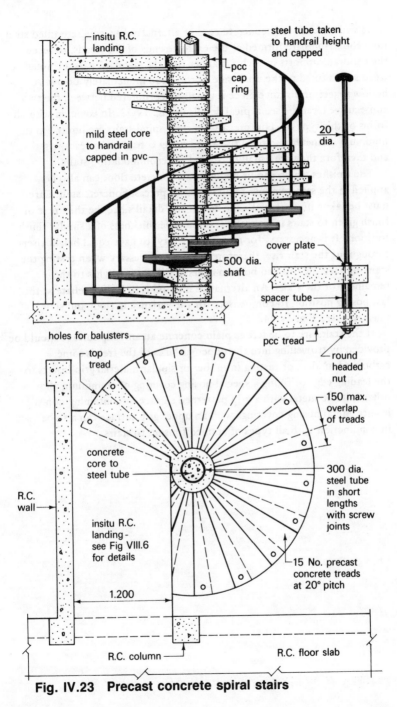

Fig. IV.23 Precast concrete spiral stairs

insitu R.C. landing

steel tube taken to handrail height and capped

pcc cap ring

mild steel core to handrail capped in pvc

20 dia.

cover plate

spacer tube

pcc tread

round headed nut

500 dia. shaft

holes for balusters

top tread

concrete core to steel tube

R.C. wall

insitu R.C. landing - see Fig VIII.6 for details

1.200

R.C. column

R.C. floor slab

150 max. overlap of treads

300 dia. steel tube in short lengths with screw joints

15 No. precast concrete treads at 20° pitch

stairs with a reinforced concrete core or alternatively a concrete-filled steel tube core. Holes are formed at the extreme ends of the treads, to receive the handrail supports in such a manner that the standard passes through a tread and is fixed to the underside of the tread immediately below. A hollow spacer or distance piece is usually incorporated between the two consecutive treads — see typical details in Fig. IV.23. In common with all forms of this type of stair, precast concrete spiral stairs are limited as to minimum diameter and total rise when being considered as escape stairs and therefore they are usually installed as accommodation stairs.

The finishes which can be applied to a concrete floor can also be applied in the same manner to an *in situ* or precast concrete stair. Care must be taken however with the design and detail since the thickness of finish given to stairs is generally less than the thickness of a similar finish to floors. It is necessary, for reasons of safety, to have equal height risers throughout the stair rise; therefore it may be necessary when casting the stairs to have the top and bottom risers of different heights to the remainder of the stairs. An alternative method is to form a rebate at the last nosing position to compensate for the variance in floor and stair finishes — see Fig. IV.14.

If the stairs are to be left as plain concrete an anti-slip surface should be provided by trowelling into the upper surfaces of the treads some carborundum dust or casting in rubber or similar material grip inserts to the leading edge or fixing a special nosing covering of aluminium alloy or other suitable metal with a grip patterned surface or containing non-slip inserts. Metal nosing coverings with an upper grip surface can also be used in conjunction with all types of applied finishes to stairs.

31
Metal stairs

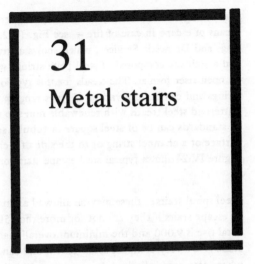

Metal stairs can be constructed to be used as escape stairs or accommodation stairs both internally and externally. Most metal stairs are manufactured from mild steel with treads of cast iron or mild steel and in straight flights with intermediate half space landings. Spiral stairs in steel are also produced but their use as an escape stair is limited by size and the number of persons likely to use the stairway in the event of a fire. Aluminium alloy stairs are also made and are used almost exclusively as internal accommodation stairs.

Since the layout of most buildings is different stairs are very often purpose made to suit the particular situation. Concrete being a flexible material at the casting stage generally presents little or no problems in this respect, whereas purpose-made metal stairs can be more expensive and take longer to fabricate in the workshop. Metal spiral stairs have the distinct advantage that the need for temporary support and hoisting equipment is eliminated. All steel stairs have the common disadvantage of requiring regular maintenance in the form of painting as a protection against corrosion.

Most metal stairs are supplied in a form which requires some site fabrication and this is usually carried out by the supplier's site erection staff, the main contractor having been supplied with the necessary data as to foundation pads, holding down bolts, any special cast-in fixings and any pockets to be left in the structural members or floor slabs to enable this preparatory work to be completed before the stairs are ready to be fixed.

Steel escape stairs: these have already been considered in the context of means of escape in case of fire — see Fig. IV.27 in Building Finishes, Fittings and Domestic Services, which shows a structural steel support frame and a stairway composed of steel plate strings with preformed treads giving an open riser format. The treads for this type of stair are bolted to the strings and can be of a variety of types ranging from perforated cast iron to patterned steel treads with renewable non-slip nosings. Handrail balustrades or standards can be of steel square or tubular sections bolted to the upper surface of a channel string or to the side of a channel or steel plate string. Figure IV.24 shows typical steel escape stair components.

Steel spiral stairs: these may be allowed as an internal or external means of escape stairs if they are not for more than 50 persons, the maximum total rise is 9 000 and the minimum overall diameter is 1.500 m. Spiral stairs give a very compact arrangement and can be the solution in situations where plan area is limited. In common with all steel external escape stairs the tread and landing plates should have a non-slip surface and be self draining with the stairway circulation width completely clear of any opening doors. Two basic forms are encountered, namely those with treads which project from the central pole or tube and those which have riser legs. The usual plan format is to have 12 or 16 treads to complete one turn around the central core and terminating at floor level with a quarter circle landing or square landing. The standards, like those used for precast concrete spiral stairs, pass through one tread and are secured on the underside of the tread immediately below, giving strength and stability to both handrail and steps. Handrails are continuous and usually convex in cross section of polished metal, painted metal or plastic covered. Typical details are shown in Fig. IV.25.

String beam steel stairs: used mainly to form accommodation stairs which need to be light and elegant in appearance; this is achieved by using small sections and an open riser format. The strings can be of mild steel tube, steel channel, steel box or small universal beam sections fixed by brackets to the upper floor surfaces or landing edges to act as inclined beams. The treads, which can be of hardwood timber, precast concrete or steel, are supported by plate, angle or tube brackets welded to the top of the string beam. Balustrading can be fixed through the ends of the treads or alternatively supported by brackets attached to the outer face of the string beam — typical details are shown in Fig. IV.26.

Pressed steel stairs: accommodation stairs made from light pressed metal such as mild steel. Each step is usually pressed as one unit with the tread

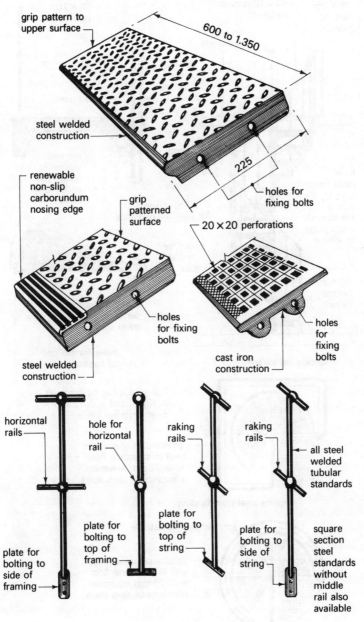

grip pattern to
upper surface

600 to 1.350

steel welded
construction

225

holes for
fixing bolts

renewable
non-slip
carborundum
nosing edge

grip
patterned
surface

20 × 20 perforations

holes
for fixing
bolts

holes
for
fixing
bolts

steel welded
construction

cast iron
construction

horizontal
rails

hole for
horizontal
rail

raking
rails

raking
rails

all steel
welded
tubular
standards

plate for
bolting to
top of
framing

plate for
bolting to
top of
string

plate for
bolting to
side of
string

square
section
steel
standards
without
middle
rail also
available

plate for
bolting to
side of
framing

For typical steel stair detail see Fig IV.27

Fig. IV.24   Typical steel stair components

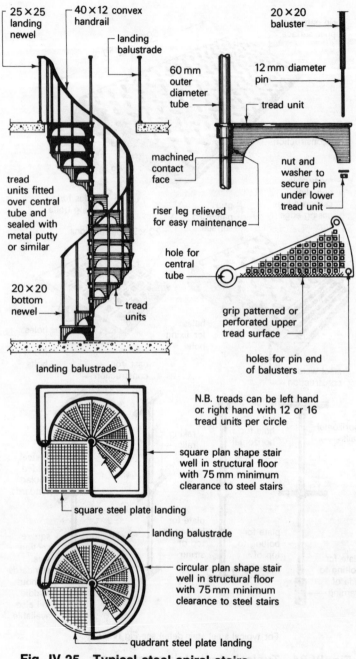

25 × 25 landing newel

40 × 12 convex handrail

landing balustrade

20 × 20 baluster

60 mm outer diameter tube

12 mm diameter pin

tread unit

tread units fitted over central tube and sealed with metal putty or similar

machined contact face

riser leg relieved for easy maintenance

nut and washer to secure pin under lower tread unit

20 × 20 bottom newel

tread units

hole for central tube

grip patterned or perforated upper tread surface

holes for pin end of balusters

landing balustrade

N.B. treads can be left hand or right hand with 12 or 16 tread units per circle

square plan shape stair well in structural floor with 75 mm minimum clearance to steel stairs

square steel plate landing

landing balustrade

circular plan shape stair well in structural floor with 75 mm minimum clearance to steel stairs

quadrant steel plate landing

Fig. IV.25   Typical steel spiral stairs

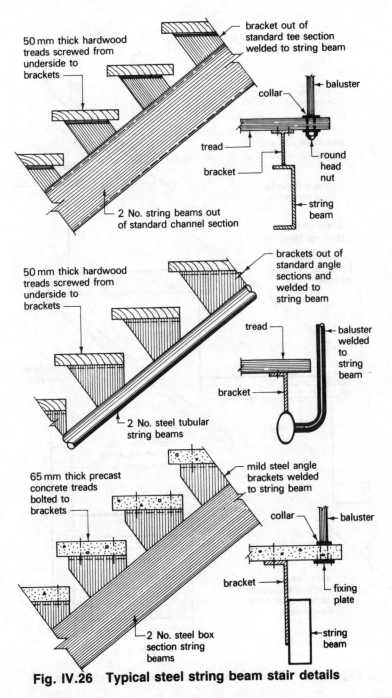

50 mm thick hardwood treads screwed from underside to brackets

bracket out of standard tee section welded to string beam

collar

baluster

tread

bracket

round head nut

string beam

2 No. string beams out of standard channel section

50 mm thick hardwood treads screwed from underside to brackets

brackets out of standard angle sections and welded to string beam

tread

baluster welded to string beam

bracket

2 No. steel tubular string beams

65 mm thick precast concrete treads bolted to brackets

mild steel angle brackets welded to string beam

collar

baluster

bracket

fixing plate

string beam

2 No. steel box section string beams

Fig. IV.26 Typical steel string beam stair details

297

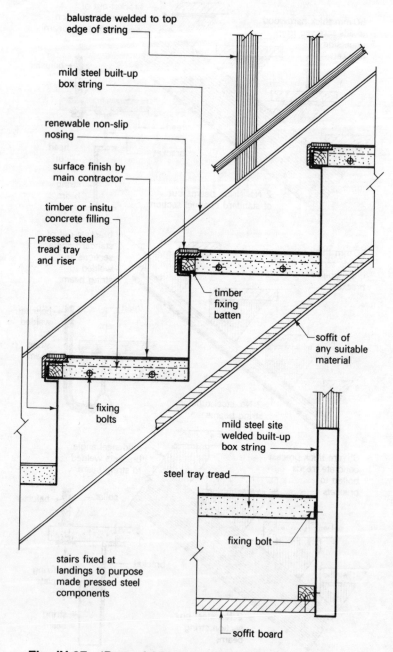

balustrade welded to top edge of string

mild steel built-up box string

renewable non-slip nosing

surface finish by main contractor

timber or insitu concrete filling

pressed steel tread tray and riser

timber fixing batten

soffit of any suitable material

fixing bolts

mild steel site welded built-up box string

steel tray tread

fixing bolt

stairs fixed at landings to purpose made pressed steel components

soffit board

Fig. IV.27 'Prestair' internal pressed steel stairs

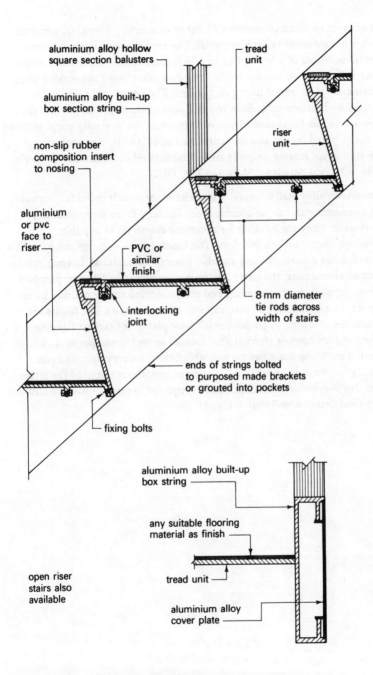

aluminium alloy hollow
square section balusters

tread
unit

aluminium alloy built-up
box section string

riser
unit

non-slip rubber
composition insert
to nosing

aluminium
or pvc
face to
riser

PVC or
similar
finish

interlocking
joint

8 mm diameter
tie rods across
width of stairs

ends of strings bolted
to purposed made brackets
or grouted into pockets

fixing bolts

aluminium alloy built-up
box string

any suitable flooring
material as finish

tread unit

aluminium alloy
cover plate

open riser
stairs also
available

Fig. IV.28 'Gradus' aluminium alloy stairs

component recessed to receive a filling of concrete, granolithic, terrazzo, timber or any other suitable material. The strings are very often in two pieces consisting of a back plate to which the steps are fixed and a cover plate to form a box section string, the cover plate being site welded using a continuous MIG (metal inert gas) process. The completed strings are secured by brackets or built in to the floors or landings and provide the support for the balustrade. Stairs of this nature are generally purpose made to the required layout and site assembled and fixed by a specialist sub-contractor leaving only the tread finishes and decoration as builder's work — typical details are shown in Fig. IV.27.

Aluminium alloy stairs: usually purpose made to suit individual layout requirements with half or quarter space landings from aluminium alloy extrusions. They are suitable for accommodation stairs in public buildings, shops, offices and flats. The treads have a non-slip nosing with a general tread covering of any suitable floor finish material. Format can be open or closed riser, the latter having greater strength. The two part box strings support the balustrading and are connected to one another by small diameter tie rods which in turn support the tread units. The flights are secured by screwing to purpose made base plates or brackets fixed to floors and landings or alternatively located in preformed pockets and grouted in. When the stairs are assembled they are very light and can usually be lifted and positioned by two men without the need for lifting gear. No decoration or maintenance is required except for routine cleaning — typical details are shown in Fig. IV.28.

Part V
Roofs

32
Roofs – timber, flat and pitched

TIMBER FLAT ROOFS

A flat roof is essentially a low pitched roof and is defined in BS 3589 as a pitch of 10° or less to the horizontal. Generally the angle of pitch is governed by the type of finish which is to be applied to the roof.

The functions of any roof are:

1. To keep out rain, wind, snow and dust.
2. To prevent excessive heat loss in winter.
3. To keep the interior of the building cool in summer.
4. Designed to accommodate all stresses encountered.
5. Designed to accept movement due to changes in temperature and moisture content.

The simplest form of roof construction to fulfil these functions is a timber flat roof covered with an impervious material to prevent rain penetration. This form of roof is suitable for spans up to 4 000 mm, spans over this are usually covered with a reinforced concrete slab or a patent form of decking.

The disadvantages of timber flat roofs are:

(a) They are poor insulators against the transfer of heat.
(b) They tend to give low rise buildings an unfinished appearance.
(c) Unless properly designed and constructed pools of water will collect on the surface causing local variations in temperature. This

results in deterioration of the covering and, consequently, high maintenance costs.

Construction

The construction of a timber flat roof follows the same methods as those employed for the construction of timber upper floors. Suitable joist sizes can be obtained by design or by reference to Tables B21 and B22 given in Approved Document A. The spacing of roof joists is controlled by the width of decking material to be used and/or the width of ceiling board on the underside. Timber flat roofs are usually constructed to fall in one direction towards a gutter or outlet. This can be achieved by sloping the joists to the required fall but, as this would give a sloping soffit on the underside, it is usual to fix wedge shaped fillets called 'firrings' to the top of the joists to provide the fall (see Fig. V.1). The materials used in timber flat roof construction are generally poor thermal insulators and therefore some form of non-structural material can be incorporated into the roof if it has to comply with Part L of the Building Regulations.

DECKING MATERIALS

Timber: This can be in the form of softwood boarding, chipboard or plywood. Plain edge sawn boards or tongued and grooved boards are suitable for joists spaced at centres up to 450 mm. Birch or fir plywood is available in sheet form which requires to be fixed on all four edges—this means that noggins will be required between joists to provide the bearing for the end fixings. Chipboard is also a sheet material and is fixed in a similar manner to plywood by using noggins and it should be noted that this material is susceptible to moisture movement. Flat roofs using a timber decking should have the roof void ventilated to minimise moisture content fluctuations; therefore it is advisable to use structural timbers which have been treated against fungal and insect attack. In certain areas treatment to prevent infestation by the house longhorn beetle is required under Building Regulation 7 (Table 1 in supporting AD).

Compressed straw slabs: these are made from selected straw by a patent method of heat and pressure to a standard width of 1 200 mm with a selection of lengths, the standard thickness being 50 mm which gives sufficient strength for the slabs to span 600 mm. All edges of the slabs must be supported and fixed. Ventilation is of the utmost importance and it is common practice to fix cross bearers at right-angles to and over the joists to give cross ventilation. A bitumen scrim should be placed over the joints before the weathering membrane is applied.

304

Wood wool slabs: these are 600 mm wide slabs of various lengths and thicknesses which can span up to 1 200 mm. The slabs are made of shredded wood fibres which have been chemically treated and are bound together with cement. The fixing and laying is similar to compressed straw slabs.

INSULATING MATERIALS

There are many types of insulating materials available, usually in the form of boards or quilts. Boards are laid over the joists, either under or on top of the rough boarding, whereas quilted materials are laid over or between the joists.

Boards: these are made from lightly compressed vegetable fibres which are bonded together with glues or resins. Being lightly compressed they contain a large number of tiny voids in which air is trapped and it is this entrapped air which makes them good thermal insulators. Since these boards are light in weight, low in strength and are compressible they must have adequate support and fixings. Fibreboards are made to a wide variety of sizes, the thicker boards being the better insulators.

Compressed straw slabs and wood wool slabs are also good thermal insulators and can be used in the dual role of decking and insulation.

Quilts: these are made from mineral or glass wool which is loosely packed between sheets of paper, since they are fine shreds giving rise to irritating scratches if handled. Quilts rely on the loose way in which the core is packed for their effectiveness and therefore the best results are obtained when they are laid between joists.

A variety of loose fills are also available for placing between the joists and over the ceiling to act as thermal insulators.

WEATHER-PROOF FINISHES

Suitable materials are asphalt, lead, copper, zinc, aluminium and built-up roofing felt; only the latter will be considered at this stage.

Built-up roofing felt
Most roofing felts consist essentially of a base sheet which is impregnated with hot bitumen during manufacture and is then coated on both sides with a stabilised weatherproof bitumen compound. This outer coating is dusted while still hot and tacky with powdered talc or receives an application of fine or coarse sand or coloured mineral granules. After

cooling the felt is cut to form rolls 1 m wide and 10 or 20 m long before being wrapped for dispatch.

BS 747 Part 2 specifies three classes of base felt:

Class 1: fibre base—this type is very flexible and of low cost.
Class 2: asbestos base—this type has good fire resistance, is fairly stable and is virtually rot proof.
Class 3: glass fibre base—this type is rot proof, very stable and is used for high quality work.

For flat roofs three layers of felt should be used the first being laid at right-angles to the fall commencing at the eaves. If the decking is timber the first layer is secured with large flat-head felt nails and the subsequent layers are bonded to it with a hot bitumen compound by a roll and pour method. If the decking is of a material other than timber all three layers are bonded with hot bitumen compound. It is usually recommended that a vented first layer is used in case moisture is trapped during construction; this recommendation does not normally apply to roofs with a timber deck since timber has the ability to 'breathe'. The minimum fall recommended for built-up roofing felt is 17 mm in 1 000 mm or 1°.

In general the Building Regulations require a flat roof with a weather-proofing which has a surface finish of asbestos based bituminous felt or is covered with a layer of stone chippings. The chippings protect the underlying felt, provide additional fire resistance and give increased solar reflection. A typical application would be 12·5 mm stone chippings at approximately 50 kilogrammes to each 2·5 square metres of roof area. Chippings of limestone, granite and light coloured gravel would be suitable.

VAPOUR BARRIERS

The problem of condensation should always be considered when constructing a flat roof. The insulation below the built-up roofing felt will not prevent condensation occurring and since it is a permeable material water vapour will pass upwards through it and condense on the underside of the roofing felt. The drops of moisture so formed will soak into the insulating material lowering its insulation value and possibly causing staining on the underside. To prevent this occurring a vapour barrier should be placed on the underside of the insulating material. A layer of roofing felt with sealed laps is suitable, an alternative being a layer of aluminium foil (many boards can be obtained with the foil already attached).

For typical timber flat roof details see Fig. V.1

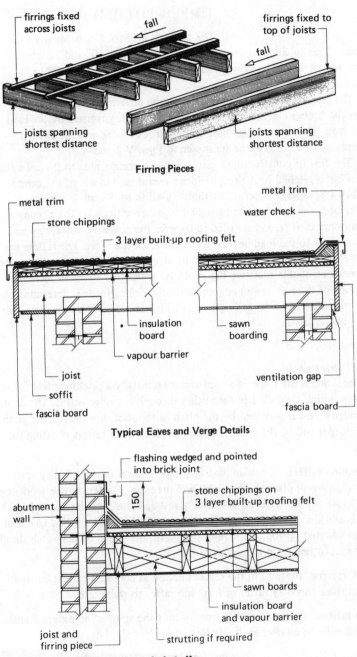

Firring Pieces

firrings fixed across joists

fall

firrings fixed to top of joists

fall

joists spanning shortest distance

joists spanning shortest distance

Typical Eaves and Verge Details

metal trim

stone chippings

3 layer built-up roofing felt

insulation board

vapour barrier

joist

soffit

fascia board

metal trim

water check

sawn boarding

ventilation gap

fascia board

flashing wedged and pointed into brick joint

150

stone chippings on 3 layer built-up roofing felt

abutment wall

sawn boards

insulation board and vapour barrier

strutting if required

joist and firring piece

Fig. V.1 Timber flat roof details

TIMBER PITCHED ROOFS

The term pitched roof includes any roof whose angle of slope to the horizontal lies between 10° and 70°; below this range it would be called a flat roof and above 70° it would be classified as a wall.

The pitch is generally determined by the covering which is to be placed over the timber carcass, whereas the basic form is governed by the load and span. The terminology used in timber roof work and the basic members for various spans are shown in Figs. V.2 and V.3. .

The cost of constructing a pitched roof is comparable to that of a flat roof but a pitched roof also provides a useful void in which to house a cold water storage cistern. The timber used in roof work is structural softwood, the members being joined together with nails. The sloping components or rafters are used to transfer the covering, wind, rain and snow loads to the load-bearing walls on which they rest. The rafters are sometimes assisted in this function by struts and purlins in what is called a purlin or double roof (see Fig. V.4). As with other forms of roofs the spacing of the rafters and consequently the ceiling joists is determined by the module size of the ceiling boards which are to be fixed on the underside of the joists.

Roof members

Ridge: this is the spine of a roof and is essentially a pitching plate for the rafters which are nailed to each other through the ridge board. The depth of ridge board is governed by the pitch of the roof, the steeper the pitch the deeper will be the vertical or plumb cuts on the rafters abutting the ridge.

Common rafters: the main load-bearing members of a roof, they span between a wall plate at eaves level and the ridge. Rafters have a tendency to thrust out the walls on which they rest and this must be resisted by the walls and the ceiling joists. Rafters are notched over and nailed to a wall plate situated on top of a load-bearing wall, the depth of the notch should not exceed one-third the depth of the rafter.

Jack rafters: these fulfil the same function as common rafters but span from ridge to valley rafter or from hip rafter to wall plate.

Hip rafters: similar to a ridge but forming the spine of an external angle and similar to a rafter spanning from ridge to wall plate.

Valley rafters: as hip rafter but forming an internal angle.

Wall plates: these provide the bearing and fixing medium for the various roof members and distribute the loads evenly over the supporting walls; they are bedded in cement mortar on top of the load-bearing walls.

Dragon ties: a tie place across the corners and over the wall plates to help provide resistance to the thrust of a hip rafter.

Ceiling joists: these fulfil the dual function of acting as ties to the feet of pairs of rafters and providing support for the ceiling boards on the underside and any cisterns housed within the roof void.

Purlins: these act as beams reducing the span of the rafters enabling an economic section to be used. If the roof has a gable end they can be supported on a corbel or built in but in a hipped roof they are mitred at the corners and act as a ring beam.

Struts: these are compression members which transfer the load of a purlin to a suitable load-bearing support within the span of the roof.

Collars: these are extra ties to give additional strength and are placed at purlin level.

Binders: these are beams used to give support to ceiling joists and counteract excessive deflections and are used if the span of the ceiling joist exceeds 2 400 mm.

Hangers: vertical members used to give support to the binders and allow an economic section to be used, they are included in the design if the span of the binder exceeds 3 600 mm.

The arrangement of struts, collars and hangers only occurs on every fourth or fifth pair of rafters.

EAVES

The eaves of a roof is the lowest edge which overhangs the wall thus giving the wall a degree of protection, it also provides the fixing medium for the rainwater gutter. The amount of projection from the wall of the eaves is a matter of choice but is generally in the region of 300-450 mm.

Two basic types of eaves are used, open eaves and closed eaves. The former is a cheap method, the rafters being left exposed on the underside and should be treated with a preservative. The space between the rafters and the roof covering is filled in with brickwork, a process called beam filling and a continuous triangular fillet is fixed over the backs of the rafters to provide the support for the bottom course of slates or tiles. The closed eaves is where the feet of the rafters are boxed in using a vertical board called a 'fascia' and the space between the fascia and the wall being filled in with a soffit board, the brick wall being terminated above the soffit level (see Fig. V.4).

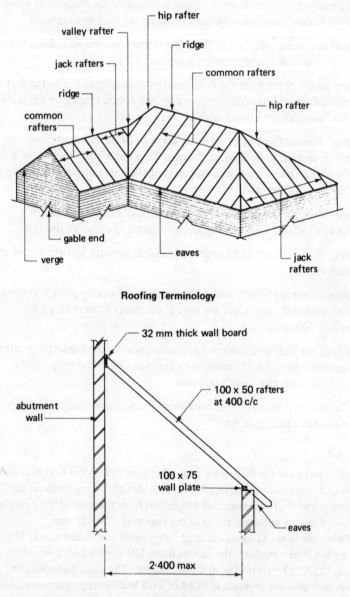

Roofing Terminology

hip rafter

valley rafter

ridge

jack rafters

common rafters

ridge

hip rafter

common rafters

gable end

verge

eaves

jack rafters

---

32 mm thick wall board

abutment wall

100 x 50 rafters at 400 c/c

100 x 75 wall plate

eaves

2·400 max

Lean-to Roof

Fig. V.2 Roofing terminology and lean-to-roof

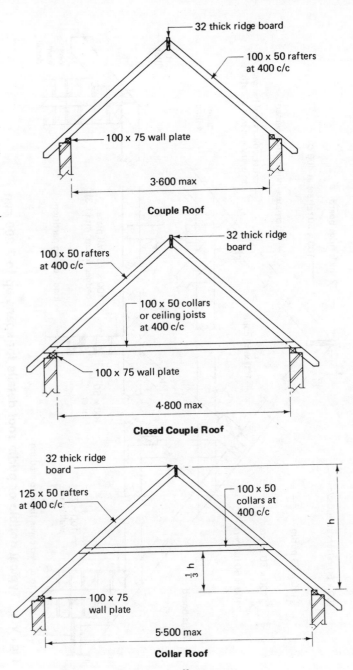

Fig. V.3 Pitched roofs for small spans

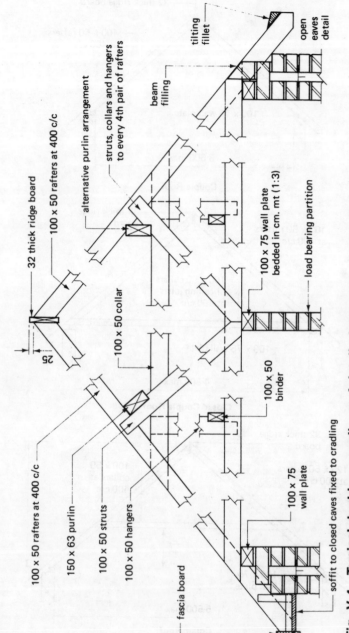

100 × 50 rafters at 400 c/c

32 thick ridge board

alternative purlin arrangement

struts, collars and hangers to every 4th pair of rafters

25

100 × 50 collar

tilting fillet

open eaves detail

beam filling

100 × 75 wall plate bedded in cm. mt (1:3)

load bearing partition

100 × 50 binder

100 × 50 rafters at 400 c/c

150 × 63 purlin

100 × 50 struts

100 × 50 hangers

100 × 75 wall plate

soffit to closed eaves fixed to cradling

fascia board

Fig. V.4 Typical double or purlin roof details for spans up to 7 200 mm

312

TIMBER ROOF TRUSSES

These can be used on the larger spans in domestic work to give an area below the ceiling level free from load-bearing walls. Trusses are structurally designed frames based on the principles of triangulation and serve to carry the purlins, they are spaced at 1 800 mm centres with the space between being infilled with common rafters. It is essential that the members of a roof truss are rigidly connected together since light sections are generally used. To make a suitable rigid joint bolts and timber connectors are used; these are square or circular toothed plates, the teeth being pointed up and down which when clamped between two members bite into the surface forming a strong connection and spread the stresses over a greater surface area. A typical roof truss detail is shown in Fig. V.5.

TRUSSED RAFTERS

This is another approach to the formation of a domestic timber roof giving a clear span; as with roof trusses it is based upon a triangulated frame but in this case the members are butt jointed and secured with truss plates. All members in a trussed rafter are machined on all faces so that they are of identical thickness ensuring a strong connection on both faces. The trussed rafters are placed at 600 mm centres and tied together over their backs with tiling battens, no purlin or ridge is required. This system of roofing is only economic if a standard span is used or if a reasonable quantity of non-standard sizes are required.

Truss or nail plates are generally of one or two forms:

1. Those in which holes are punched to take nails and are suitable for site assembly using a nailing gun.
2. Those in which teeth are punched and bent from the plate and are used in factory assembly using heavy presses.

In all cases truss plates are fixed to both faces of the butt joint.

Trussed rafters are also produced using gusset plates of plywood at the butt joints instead of truss plates, typical details of both forms are shown in Fig. V.6.

BUILDING REGULATIONS 1985

Regulation 7 requires that any building work shall be carried out with proper materials and in a workman-like manner. The Approved Document supporting this regulation defines acceptable levels of performance for materials which include products, fittings, items of equipment and back-

313

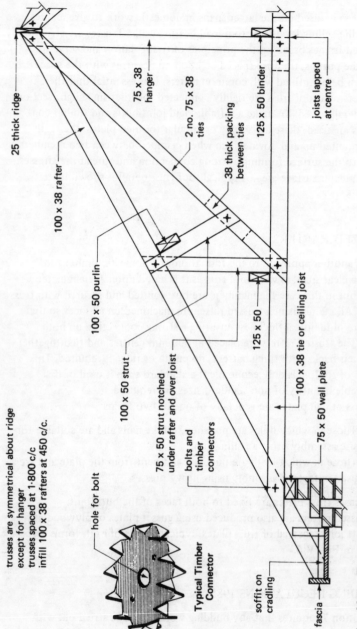

trusses are symmetrical about ridge
except for hanger
trusses spaced at 1·800 c/c
infill 100 x 38 rafters at 450 c/c.

hole for bolt

Typical Timber Connector

25 thick ridge

100 x 38 rafter

100 x 50 purlin

100 x 50 strut

75 x 50 strut notched
under rafter and over joist

bolts and timber
connectors

soffit on
cradling

fascia

75 x 38 hanger

2 no. 75 x 38
ties

38 thick packing
between ties

125 x 50 binder

joists lapped
at centre

125 x 50

100 x 38 tie or ceiling joist

75 x 50 wall plate

Fig. V.5   Typical truss detail for spans up to 8 000 mm

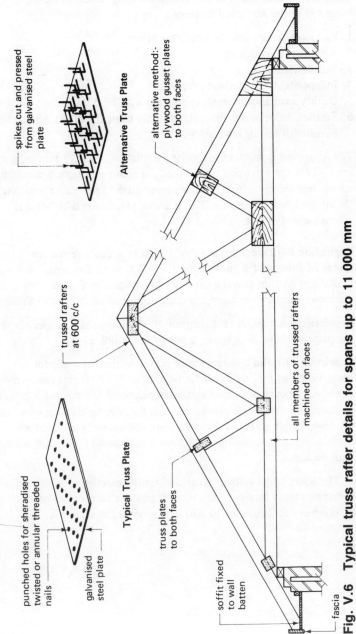

spikes cut and pressed from galvanised steel plate

Alternative Truss Plate

alternative method:- plywood gusset plates to both faces

trussed rafters at 600 c/c

all members of trussed rafters machined on faces

punched holes for sheradised twisted or annular threaded nails

galvanised steel plate

Typical Truss Plate

truss plates to both faces

soffit fixed to wall batten

fascia

Fig. V.6 Typical truss rafter details for spans up to 11 000 mm

filling for excavations. The aids for establishing fitness of materials suggested in the Approved Document are:

1. Past experience—such as a building in use.
2. Agrément Certificates.
3. British Standards.
4. Independent certification schemes.
5. Quality assurance schemes complying with BS 5750.
6. Test and calculations for materials only using the NATLAS accreditation scheme for testing laboratories.

The Approved Document supporting regulation 7 lists certain geographical areas where the softwood timber used for roof construction should be adequately treated with a suitable preservative to prevent infestation by the House Longhorn Beetle. This is a wise precaution whether it is recommended or not.

Regulation B4(2) requires the roof to offer adequate resistance to the spread of fire over the roof. Table 1.3 in Approved Document B gives limitations on roof coverings for dwelling houses by designations and distance from the boundary, the designations being defined in Table A4.

Regulation C4 requires that the roof of a building shall adequately resist the passage of moisture to the inside of the building.

Regulation L1 states that reasonable provision shall be made for the conservation of fuel and power in buildings. To satisfy this requirement Approved Document L gives maximum permitted U values of $0.25 \text{W}/\text{m}^2\text{K}$ for dwellings with single glazing and $0.35 \text{W}/\text{m}^2\text{K}$ for dwellings with all windows double glazed. These values are not normally achieved with traditional roof constructions and therefore a suitable insulating material must be added.

The selection of suitable structural timber members for pitched roofs for single family houses of not more than three storeys can be ascertained by reference to Tables B5 to B20 in Approved Document A.

33
Roof tiling and slating

A roof can be defined as the upper covering of a building and in this position it is fully exposed to the rain, snow, wind, sun and general atmosphere, therefore the covering to any roof structure must have good durability to meet these conditions. Other factors to be taken into account are weight, maintenance and cost.

Roofs are subjected to wind pressures, both positive and negative, the latter causing uplift and suction which can be overcome by firmly anchoring lightweight coverings to the structure or by relying upon the deadweight of the covering material. Domestic roofs are not usually designed for general access and therefore the chosen covering material should be either very durable or easily replaceable. The total dead load of the covering will affect the type of support structure required and ultimately the total load on the foundations, so therefore careful consideration must be given to the medium selected for the roof covering.

In domestic work the roof covering usually takes the form of tiling or slating since these materials economically fulfil the above requirements and have withstood the test of time.

Tiling

Tiles are manufactured from clay and concrete to a wide range of designs and colours suitable for pitches from 17-45° and work upon the principle of either double or single lap. The vital factor for the efficient performance of any tile or slate is the pitch and it should be noted that the pitch of a tile is always less than the pitch of the rafters owing to the overlapping technique.

Tiles are laid in overlapping courses and rely upon the water being shed off the surface of one tile onto the exposed surface of the tile in the next course. The problem of water entering by capillary action between the tiles is overcome by the camber of the tile, the method of laying or by overlapping side joints. In all methods of tiling a wide range of fittings are produced to enable all roof shapes to be adequately protected.

PLAIN TILING

This is a common method in this country and works on the double lap principle. The tiles can be of hand-made or machine pressed clay, the process of manufacture being similar to that of brickmaking. The hand-made tile is used mainly where a rustic or distinctive roof character is required since they have a wide variation in colour, texture and shape. They should not be laid on a roof of less than 45° pitch since they tend to absorb water and if allowed to become saturated they may freeze, expand and spall or fracture in cold weather. Machine pressed tiles are harder, denser and more uniform in shape than hand-made varieties and can be laid to a minimum pitch of 35°.

A suitable substitute for plain clay tiles is the concrete plain tile, these are produced in a range of colours to the same size specifications as the clay tiles and with the same range of fittings. The main advantage of concrete tiles is the lower cost and the main disadvantage is the extra weight. Plain clay tiles are covered by BS 402 and concrete tiles are covered by BS 473 and 550.

Plain tiling in common with other forms of tiling provides an effective barrier to rain and snow penetration but wind is able to penetrate into the building through the gaps between the tiling units, therefore a barrier in the form of boarding or sheeting is placed over the roof carcass before the battens on which the tiles are to be hung are fixed.

The rule for plain tiling is that there must always be at least two thicknesses of tiles covering any part of the roof and bonded so that no 'vertical' joint is immediately over a 'vertical' joint in the course below. To enable this rule to be maintained shorter length tiles are required at the eaves and the ridge, each alternate course is commenced with a wider tile of one-and-a-half tile widths. The apex or ridge is capped with a special tile bedded in cement mortar over the general tile surface. The hips can be covered with a ridge tile in which case the plain tiling is laid underneath and mitred on top of the hip or alternatively a special bonnet tile can be used where the plain tiles bond with the edges of the bonnet tiles. Valleys can be formed by using special tiles, mitred plain tiles or forming an open

318

gutter with a durable material such as lead. The verge at the gable end can be formed by bedding plain tiles face down on the gable wall as an undercloak and bedding the plain tiles in cement mortar on the upper surface of the undercloak. The verge tiling should overhang its support by at least 50 mm. Abutments are made watertight by dressing a flashing over the upper surface of tiling between which is sandwiched a soaker. The soaker in effect forms a gutter. An alternative method is to form a cement fillet on top of the tiled surface but this method sometimes fails by the cement shrinking away from the surface of the wall.

The support or fixing battens are of softwood extended over and fixed to at least three rafters, the spacing or gauge being determined by the lap given to the tiles thus:

$$\text{Gauge} = \frac{\text{length of tile} - \text{lap}}{2}$$

$$\text{Gauge} = \frac{265 - 65}{2} = 100 \text{ mm.}$$

Plain tiles are fixed with two galvanised nails to each tile in every fourth or fifth course.

Details of plain tiles, fittings and methods of laying are shown in Figs. V.7 to V.11.

SINGLE LAP TILING

Single lap tiles are laid with overlapping side joints to a minimum pitch of 35° and are not bonded like the butt jointed single lap plain tiles; this gives an overall reduction in weight since less tiles are used. A common form of single lap tile is the pantile which has opposite corners mitred to overcome the problem of four tile thicknesses at the corners (see Fig. V.12). The pantile is a larger unit than the plain tile and is best employed on large roofs with gabled ends since the formation of hips and valleys is difficult and expensive. Other forms of single lap tiling are Roman tiling, Spanish tiling and interlocking tiling. The latter types are produced in both concrete and clay and have one or two grooves in the overlapping edge to give greater resistance to wind penetration and can generally be laid to lower pitches than other forms of tiling (see Fig. V.12).

Slating

Slate is a naturally dense material which can be split into thin sheets and used to provide a suitable covering to a pitched roof. Slates are laid to the same basic principles as double lap

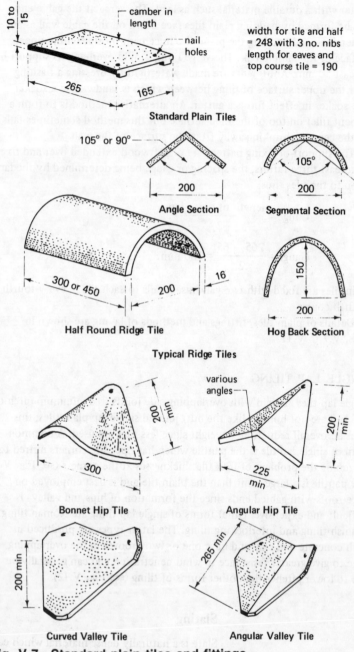

Fig. V.7 Standard plain tiles and fittings

Labels within figure:

10 to 15

camber in length

nail holes

265

165

width for tile and half = 248 with 3 no. nibs length for eaves and top course tile = 190

Standard Plain Tiles

105° or 90°

200

Angle Section

105°

200

Segmental Section

300 or 450

200

16

Half Round Ridge Tile

150

200

Hog Back Section

Typical Ridge Tiles

200 min

300

Bonnet Hip Tile

various angles

200 min

225 min

Angular Hip Tile

200 min

Curved Valley Tile

265 min

Angular Valley Tile

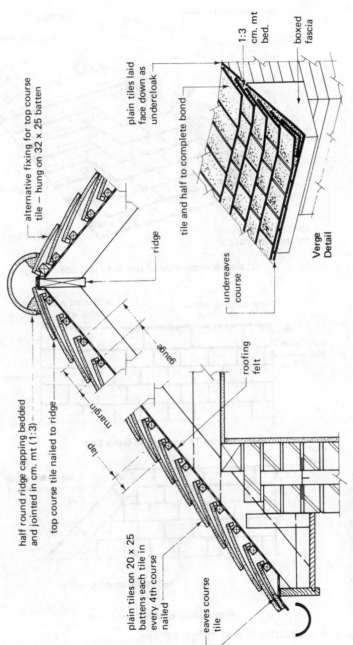

alternative fixing for top course tile – hung on 32 x 25 batten

half round ridge capping bedded and jointed in cm. mt (1:3)

top course tile nailed to ridge

ridge

gauge

margin

lap

plain tiles on 20 x 25 battens each tile in every 4th course nailed

eaves course tile

roofing felt

1:3 cm. mt bed.

boxed fascia

plain tiles laid face down as undercloak

tile and half to complete bond

undereaves course

Verge Detail

Fig. V.8 Plain tiling details

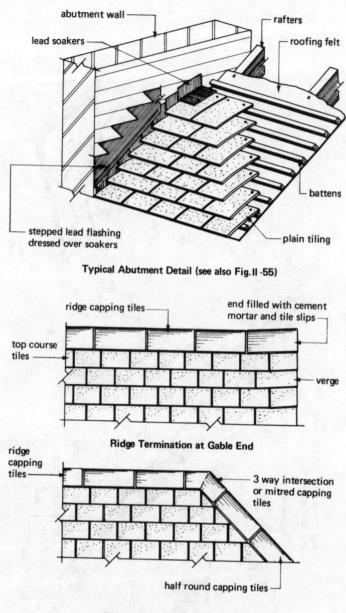

Typical Abutment Detail (see also Fig. II-55)

Ridge Termination at Gable End

Ridge Junction with Hipped End

Fig. V.9 Abutment and ridge details

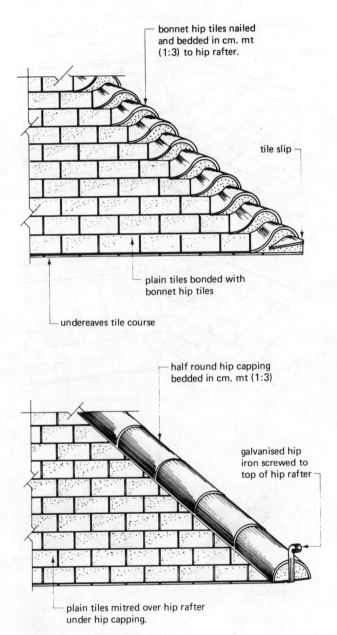

bonnet hip tiles nailed
and bedded in cm. mt
(1:3) to hip rafter.

tile slip

plain tiles bonded with
bonnet hip tiles

undereaves tile course

half round hip capping
bedded in cm. mt (1:3)

galvanised hip
iron screwed to
top of hip rafter

plain tiles mitred over hip rafter
under hip capping.

Fig. V.10 Hip treatments

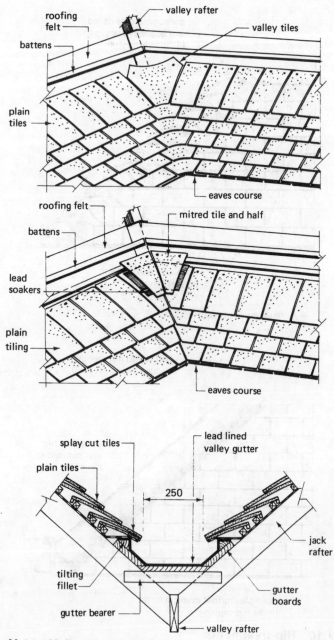

Fig. V.11 Valley treatments

roofing tiles except that every slate should be twice nailed. Slates come mainly from Wales, Cornwall and the Lake District and are cut to a wide variety of sizes—the Westmorland slates are harder to cut and are usually supplied in random sizes. Slates can be laid to a minimum pitch of $25°$ and are fixed by head nailing or centre nailing. Centre nailing is used to overcome the problem of vibration caused by the wind and tending to snap the slate at the fixing if nailed at the head, it is used mainly on the long slates and pitches below $35°$.

The gauge of the battens is calculated thus:

$$\text{Head nailed gauge} = \frac{\text{length of slate} - (\text{lap} + 25 \text{ mm})}{2}$$

$$= \frac{400 - (75 + 25)}{2} = 150 \text{ mm}$$

$$\text{Centre nailed gauge} = \frac{\text{length of slate} - \text{lap}}{2}$$

$$= \frac{400 - 76}{2} = 162 \text{ mm}.$$

Roofing slates are covered by BS 680 which gives details of standard sizes, thicknesses and quality. Typical details of slating are shown in Fig. V.13.

Roofing felts

The object of a roofing felt is to keep out dust and the wind, also to assist in providing an acceptable level of thermal and sound insulation. It consists basically of a bituminous impregnated matted fibre sheet which can be reinforced with a layer of jute hessian embedded in the coating on one side to overcome the tendency of felts to tear readily. Roofing felts are supplied in rolls 1 m wide and 10 or 20 m long depending upon type. They should be laid over the rafters and parallel to the eaves with 150 mm laps and temporarily fixed with large head felt nails until finally secured by the battens. Roofing felts should conform to the requirements of BS 747.

Counter battens

If a roof is boarded before the roofing felt is applied the tiling battens will provide ledges on which dirt and damp can collect. To overcome this problem counter battens are fixed to the boarding over the rafter positions to form a cavity between the tiling battens and the boarding. Boarding a roof over the rafters is a method of providing a wind barrier and adding to the thermal insulation properties but is seldom used in new work because of the high cost.

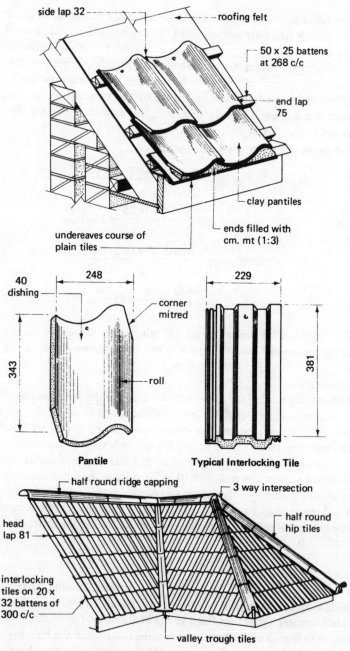

Fig. V.12 Examples of single lap tiling

- side lap 32
- roofing felt
- 50 x 25 battens at 268 c/c
- end lap 75
- clay pantiles
- ends filled with cm. mt (1:3)
- undereaves course of plain tiles
- 40 dishing
- 248
- corner mitred
- 343
- roll
- **Pantile**
- 229
- 381
- **Typical Interlocking Tile**
- half round ridge capping
- 3 way intersection
- head lap 81
- half round hip tiles
- interlocking tiles on 20 x 32 battens of 300 c/c
- valley trough tiles

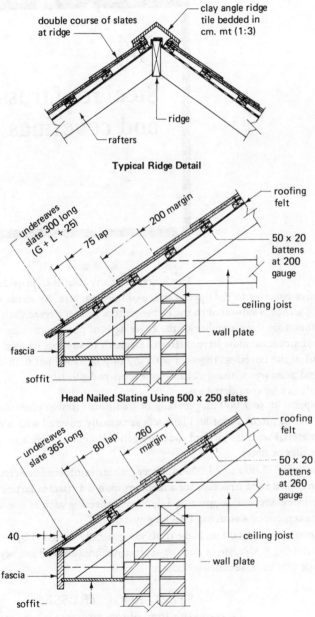

Typical Ridge Detail

double course of slates at ridge

clay angle ridge tile bedded in cm. mt (1:3)

ridge

rafters

Head Nailed Slating Using 500 x 250 slates

undereaves slate 300 long (G + L + 25)

75 lap

200 margin

roofing felt

50 x 20 battens at 200 gauge

ceiling joist

wall plate

fascia

soffit

Centre Nailed Slating Using 600 x 300 slates

undereaves slate 365 long

80 lap

260 margin

roofing felt

50 x 20 battens at 260 gauge

ceiling joist

wall plate

40

fascia

soffit

Fig. V.13 Typical slating details

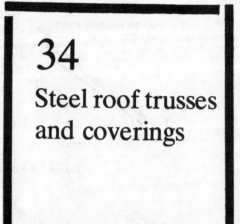

34
Steel roof trusses and coverings

The function of any roof is to provide a protective covering to the upper surface of the structure. By virtue of its position a roof is subjected to the elements to a greater degree than the walls, therefore the durability of the covering is of paramount importance. The roof structure must have sufficient strength to support its own weight, the load of the coverings together with any imposed loadings such as snow and wind pressures without collapse or excessive deflection.

Roofs can be considered as:

Short span: up to 7.000 m, generally of traditional timber construction with a flat or pitched profile. Flat roofs are usually covered with a flexible sheet material whereas pitched roofs are generally covered with small units such as tiles or slates.

Medium span: 7.000 to 24.000 m, except where reinforced concrete is used; the usual roof structure for a medium span is a truss or lattice of standard steel sections supporting a deformed sheeting such as corrugated asbestos cement or a structural decking system.

Long span: over 24.000 m; these roofs are generally designed by a specialist using girder, space deck or vaulting techniques and are beyond the scope of a basic technology course.

STEEL ROOF TRUSSES

This form of roof structure is used mainly for short and medium span single storey buildings intended for industrial or recreational use. A steel roof truss is a plane frame consisting of a series of rigid triangles composed of compression and tension members. The

328

compression members are called rafters and struts, whereas the tension members are termed ties. Standard mild steel angles complying with the recommendations of BS 4848 are usually employed as the structural members and these are connected together, where the centre lines converge, with flat shaped plates called gussets. They can be rivetted, bolted or welded together to form a rigid triangulated truss; typical arrangements are shown in Fig. V.14. The internal arrangement of the struts and ties will be governed by the span. The principal or rafter is divided into a number of equal divisions which locates the intersection point for the centre line of the internal strut or tie.

Angle purlins are used longitudinally to connect the trusses together and provide the fixing medium for the roof covering. It is the type of covering chosen which will determine the purlin spacing and the pitch of the truss; ideally the purlins should be positioned over the strut or tie intersection points to avoid setting up local bending stresses in the rafters. Purlins are connected to cleats attached to the backs of the rafters; alternatively a zed section can be used, thus dispensing with the need for a fixing cleat. Steel roof trusses are positioned at 3.000 to 7.500 m centres and supported by capped universal columns or bolted to padstones bedded on to brick walls or piers. The main disadvantage of this form of roofing is the large and virtually unusable roof space. Other disadvantages are the frequent necessity of painting the members to inhibit rust and that the flanges of the angles provide an ideal ledge on which dust can accumulate. Typical steel roof truss details are shown in Fig. V.15.

Suitable truss and girder arrangements can be fabricated from welded steel tubes which are lighter in weight, cleaner in appearance, have less surface area on which to collect dust and therefore less surface area to protect with paint.

Coverings

The basic requirements for covering materials to steel roof trusses are:

1. Sufficient strength to support imposed wind and snow loadings.
2. Resistance to the penetration of rain, wind and snow.
3. Low self weight, so that supporting members of an economic size can be used.
4. Reasonable standard of thermal insulation.
5. Acceptable fire resistance.
6. Durable to reduce the maintenance required during the anticipated life of the roof.

Most of the materials used for covering a steel roof structure have poor thermal insulation properties and therefore a combination of materials is

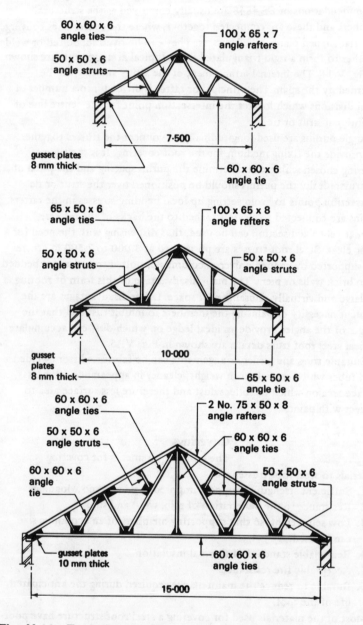

Fig. V.14 Typical mild steel angle roof trusses

60 x 60 x 6
angle ties

100 x 65 x 7
angle rafters

50 x 50 x 6
angle struts

7·500

gusset plates
8 mm thick

60 x 60 x 6
angle tie

65 x 50 x 6
angle ties

100 x 65 x 8
angle rafters

50 x 50 x 6
angle struts

50 x 50 x 6
angle struts

10·000

gusset
plates
8 mm thick

65 x 50 x 6
angle tie

60 x 60 x 6
angle ties

2 No. 75 x 50 x 8
angle rafters

50 x 50 x 6
angle struts

60 x 60 x 6
angle ties

60 x 60 x 6
angle
tie

50 x 50 x 6
angle struts

gusset plates
10 mm thick

60 x 60 x 6
angle ties

15·000

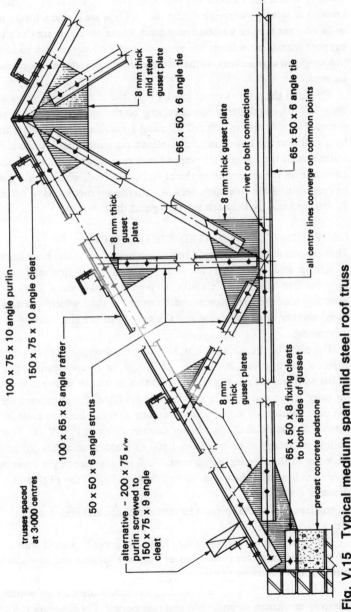

trusses spaced
at 3·000 centres

100 x 75 x 10 angle purlin

150 x 75 x 10 angle cleat

100 x 65 x 8 angle struts

50 x 50 x 6 angle s/w

alternative ~ 200 x 75
purlin screwed to
150 x 75 x 9 angle
cleat

8 mm thick
gusset
plate

8 mm thick mild
steel gusset plate

65 x 50 x 6 angle tie

8 mm thick gusset plate

rivet or bolt connections

65 x 50 x 6 angle tie

all centre lines converge on common points

8 mm thick gusset plates

65 x 50 x 8 fixing cleats
to both sides of gusset

precast concrete padstone

Fig. V.15 Typical medium span mild steel roof truss

required if heat loss, or gain, is to be controlled to satisfy legal requirements or merely to conserve the fuel required to heat the building.

Corrugated roofing materials, if correctly applied, will provide a covering which will exclude the rain and snow but will allow a small amount of wind penetration unless the end laps are sealed with 25 mm wide asbestos tape or a ribbon of mastic. These coverings are designed to support normal snow loads but are not usually strong enough to support the weight of operatives and therefore a crawling ladder or board should be used.

Owing to the poor thermal insulation value of these roofing materials there is a risk of condensation occurring on the internal surface of the sheets. This risk can be reduced by using a suitable lining at rafter level or by a ceiling at the lower tie level. Unless a vapour barrier is included on the warm side of the lining water vapour may pass through the lining and condense on the underside of the covering material, which could give rise to corrosion of the steel members. An alternative method is to form a 25 mm wide cavity between the lining and the covering.

GALVANISED CORRUGATED STEEL SHEETS

These sheets are often referred to as 'corrugated iron' and have been widely used for many years for small industrial and agricultural buildings; they can also be used as a cladding to post and rail fencing. They are generally made to the recommendations of BS 3083 which specifies the sizes, number of corrugations and the quality of the zinc coating or galvanising.

The pitch of the corrugations, which is the distance between centres of adjacent corrugations, is 75 mm with 7, 8, 9, or 10 corrugations per sheet width with lengths ranging from 1.500 to 3.600 m. A wide range of fittings for ridge, eaves and verge treatments are available. The sheets are secured to purlins with hook bolts, drive screws or nuts and bolts in a similar manner to that detailed for asbestos cement sheets in Fig. V.18. The purlins are spaced at centres from 1.500 to 3.000 m according to the thickness of the sheeting being used. To form a weather-tight covering the sheets are lapped at their ends and sides according to the pitch and exposure conditions:

end laps: up to 20° pitch 150 mm minimum and sealed with a bituminous mastic;

side laps: formed on edge away from the prevailing wind with a 1½ corrugation lap for conditions of normal exposure and two corrugation lap for conditions of severe exposure.

When fixed, galvanised corrugated steel sheets form a roof which is cheap to construct, strong, rigid and non-porous. On exposure the

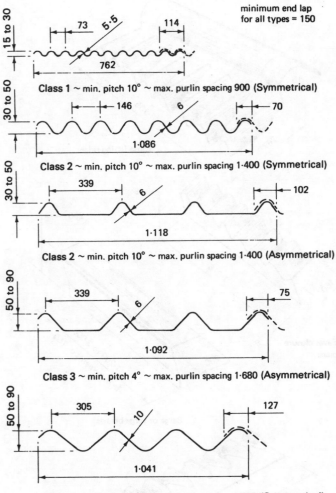

Class 1 ~ min. pitch 10° ~ max. purlin spacing 900 (Symmetrical)

Class 2 ~ min. pitch 10° ~ max. purlin spacing 1·400 (Symmetrical)

Class 2 ~ min. pitch 10° ~ max. purlin spacing 1·400 (Asymmetrical)

Class 3 ~ min. pitch 4° ~ max. purlin spacing 1·680 (Asymmetrical)

Class 4 ~ min. pitch 4° ~ max. purlin spacing 1·980 (Symmetrical)

minimum end lap
for all types = 150

Lengths:
Class 1   1 225 to 3 050
Class 2   1 525 to 3 050
Class 3   1 675 to 3 050
Class 4   1 825 to 3 050

Fig. V.16   Typical corrugated sheet profiles

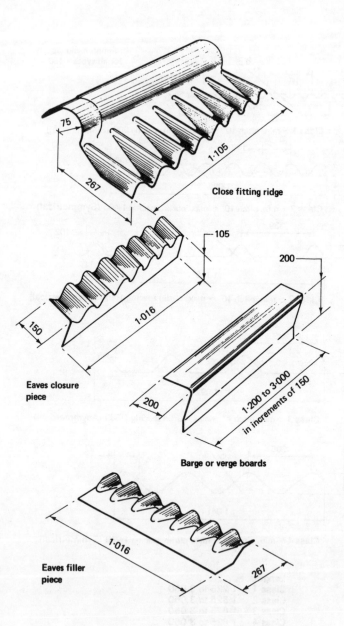

75
1·105
267
Close fitting ridge

105
1·016
150
Eaves closure
piece

200
200
1·200 to 3·000
in increments of 150
Barge or verge boards

1·016
267
Eaves filler
piece

Fig. V.17   Typical fittings for Class 2 sheeting

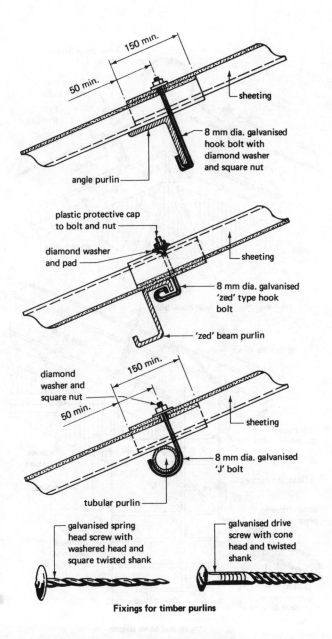

150 min.

50 min.

sheeting

8 mm dia. galvanised
hook bolt with
diamond washer
and square nut

angle purlin

plastic protective cap
to bolt and nut

diamond washer
and pad

sheeting

8 mm dia. galvanised
'zed' type hook
bolt

'zed' beam purlin

diamond
washer and
square nut

150 min.

50 min.

sheeting

8 mm dia. galvanised
'J' bolt

tubular purlin

galvanised spring
head screw with
washered head and
square twisted shank

galvanised drive
screw with cone
head and twisted
shank

Fixings for timber purlins

Fig. V.18 Typical roof sheeting fixings

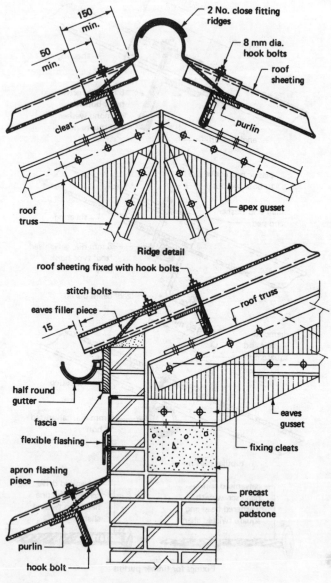

Ridge detail

- 2 No. close fitting ridges
- 8 mm dia. hook bolts
- roof sheeting
- purlin
- apex gusset
- roof truss
- cleat
- 150 min.
- 50 min.

- roof sheeting fixed with hook bolts
- stitch bolts
- eaves filler piece
- roof truss
- 15
- half round gutter
- fascia
- flexible flashing
- eaves gusset
- fixing cleats
- apron flashing piece
- precast concrete padstone
- purlin
- hook bolt

Abutment and eaves details

Fiv. V.19   Typical corrugated sheet roofing details

galvanised coating oxidises, forming a thin protective film which is easily broken down by acids in the atmosphere. To extend the life of the sheeting it should be regularly coated with paint containing a pigment of zinc dust, zinc oxide, calcium plumbate or zinc chromate. The use of these paints will eliminate the need for an application of mordant solution to provide a key. When laying new sheeting it is advisable to paint under the laps before fixing since the overlap is very vulnerable to early corrosion.

The main disadvantages of this form of roof covering are:

1. Poor thermal insulation properties — 8.6 W/m^2 °C which can be reduced by using a 12 mm insulation fibre board in conjunction with a 25 mm cavity to 1.7 W/m^2 °C.
2. Although a non-combustible material, galvanised corrugated steel sheets tend to buckle under typical fire conditions.
3. Inclined to be noisy during rain which produces a 'drumming' sound.

CORRUGATED ASBESTOS CEMENT SHEETS

This was the major covering material used for cladding steel roof structures and is made from asbestos fibres and cement in the approximate proportions of 1 : 7 together with a controlled amount of water. Corrugated asbestos cement sheets are produced to the recommendations of BS 690: Part 3 together with a wide range of fittings for the ridge, eaves and verge which are used in conjunction with the various profiles produced (see Figs. V.16 to V.19.

Concern with the health risk attached to the manufacture and use of asbestos-based products has led to the development and production of alternative fibre-based materials including profiles to match the corrugated asbestos cement sheets conforming to BS 690:Part 3.

FIBRE CEMENT PROFILED SHEETS

This alternative material to asbestos cement has been developed to meet the same technical specification with a similar low maintenance performance. Fibre cement sheets are made by combining natural and synthetic non-toxic fibres and fillers with Portland Cement and, unlike asbestos cement sheets which are rolled to form the required profile, these sheets are pressed over templates. The finished product has a natural grey colour but sheets with factory applied surface coatings are available. No British Standard yet exists for fibre cement sheeting but some products are supported by an Agrément Certificate, therefore for details of the properties, spanning ability, roof pitches and dimensions the individual manufacturer's data should be consulted.

The sheets and fittings are fixed through the crown of the corrugation using either shaped bolts to steel purlins or drive screws to timber purlins. At least six fixings should be used for each sheet and to ensure that a weather-tight seal is achieved at the fixing positions a suitable felt or lead pad with a diamond shaped curved washer can be used. Alternatively a plastic sealing washer can be employed (see Fig. V.18). The sheets can be easily drilled for fixings which should be 2 mm larger in diameter than the fixing and sited at least 40 mm from the edge of the sheet. Side laps should be positioned away from the prevailing wind and end laps on low pitches should be sealed with a mastic or suitable preformed compressible strip.

The 'U' value of fibre cement sheets is high, generally about 6.0 W/m^2 K, therefore if a higher degree of thermal resistance is required it will be necessary to use a system of underlining sheets with an insulating material sandwiched between the profile and underlining sheet.

ALUMINIUM SHEETING

This form of roof covering is available in a corrugated or troughed profile usually conforming to the requirements of BS 4868. The sheets are normally made from an aluminium-manganese alloy resulting in a non-corrosive, non-combustible lightweight sheet (2.4 to 5.0 kg/m^2). Aluminium sheeting oxides on the surface to form a protective film upon exposure to the atmosphere and therefore protective treatments are not normally necessary. Fixings of copper or brass should not be used since the electrolytic action between the dissimilar materials could cause harmful corrosion, and where the sheets are in contact with steelwork the steel members should be painted with at least two coats of zinc chromate or bituminous paint.

The general application in design and construction of an aluminium covering is similar to that described and detailed for asbestos cement sheeting. A wide range of fittings are available and like the asbestos cement sheets can be fixed with hook bolts, bolts and clips or special shot fasteners. The sheets are intended for a 15° pitched roof with purlins at 1.200 m centres for the 75 mm corrugated profile and at 2.700 m centres for the trough profiles. Laps should be 1½ corrugations for the side lap or 45—57 mm for trough sheets with a 150 mm minimum end lap for all profiles.

338

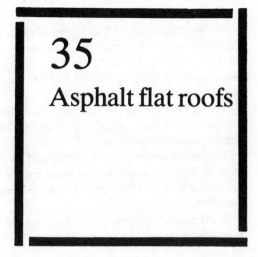

35
Asphalt flat roofs

Flat roofs are often considered to be a
simple form of construction, but unless correctly designed and constructed
they can be an endless source of trouble. Flat roofs can have the advantage
of providing an extra area to a building for the purposes of recreation and
an additional viewpoint. Mastic asphalt provides an ideal covering material
especially where foot traffic is required.

Mastic asphalt consists of an aggregate with a bituminous binder which
is cast into blocks ready for reheating on site. The blocks are heated in
cauldrons or cookers to a temperature of over 200° C and are then
transported in a liquid state in buckets for application to the roof deck by
hand spreading. Once the melted asphalt has been removed from the source
of heat it will cool and solidify rapidly, therefore the distance between the
cauldron and the point of application should be kept to a minimum. Mastic
asphalt can be applied to most types of rigid sub-structure and proprietary
structural deckings. To prevent slight movements occurring between the
sub-structure and the finish an isolating membrane should be laid over the
base before the mastic asphalt is applied.

Roof work will often entail laying mastic asphalt to horizontal, sloping
and vertical surfaces and these are defined as follows:
1. Horizontal surfaces — up to $10°$ pitch.
2. Sloping surfaces — between $10°$ and $45°$ pitch.
3. Vertical surfaces — over $45°$ pitch.
The thickness and number of coats required will depend on two factors:
1. Surface type.

2. Sub-structure or base.

Horizontal surfaces with any form of rigid base should have a two-coat application of mastic asphalt laid breaking the joint and built up to a minimum total thickness of 20 mm. An isolating membrane of black sheathing felt complying with BS 747 4A(i) should be used as an underlay laid loose with 50 mm lapped joints.

Vertical and sloping surfaces other than those with a timber base, require a three-coat application built up to a 20 mm total thickness without an isolating membrane.

Timber sub-structures with vertical and sloping surfaces should have a three-coat 20 mm mastic asphalt finish applied to expanded metal lathing complying with BS 1369 fixed at 150 mm centres over an isolating membrane.

If the mastic asphalt surface is intended for foot traffic the total thickness of the two-coat application should be increased to a minimum of 25 mm.

Roofs with a mastic asphalt finish can be laid to falls so that the run off of water is rapid and efficient. Puddles of water left on the roof surface will create small areas of different surface temperatures which could give rise to local differential thermal movements and cause cracking of the protective covering. Depressions will also provide points at which dust and rubbish can collect; it is therefore essential that suitable falls are provided to direct the water to an eaves gutter or specific outlet points. The falls should be formed on the base or supporting structure to a gradient of not less than 1 in 80 (see typical details on Figs. V.20 and V.21).

Alternatively the roof can be designed to act as a reservoir by a technique sometimes called 'ponding'. The principle is to retain on the roof surface a 'pond' of water some 150 mm deep by having the surface completely flat, high skirtings and outlets positioned 150 mm above the roof level. The main advantage is that differential temperatures are reduced to a minimum; the disadvantages are the need for a stronger supporting structure to carry the increased load, a three-coat 30 mm thick covering of mastic asphalt and the need to flood the roof in dry hot weather to prevent the pond completely evaporating away.

Thermal insulation can be provided by including into the design a dry insulation such as cork slabs, wood fibre boards and glass fibre boards. The insulation may be placed above or below the structural roof; if fixed over the roof structure it will reduce the temperature variations within the roof to a minimum and hence the risk of unacceptable thermal movements. An alternative method is to use a lightweight concrete screed to provide the falls and the thermal insulation. Suitable aggregates are furnace clinker, foamed blast furnace slag, pumice, expanded clay, sintered pulverised fuel

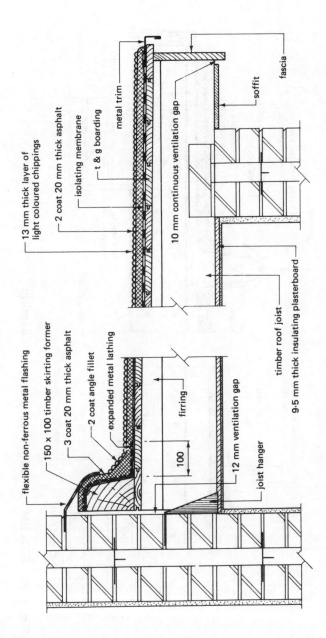

flexible non-ferrous metal flashing

150 × 100 timber skirting former

3 coat 20 mm thick asphalt

2 coat angle fillet

expanded metal lathing

firring

100

12 mm ventilation gap

joist hanger

13 mm thick layer of light coloured chippings

2 coat 20 mm thick asphalt

isolating membrane

t & g boarding

metal trim

10 mm continuous ventilation gap

soffit

fascia

timber roof joist

9·5 mm thick insulating plasterboard

Fig. V.20   Timber flat roof with mastic asphalt covering

341

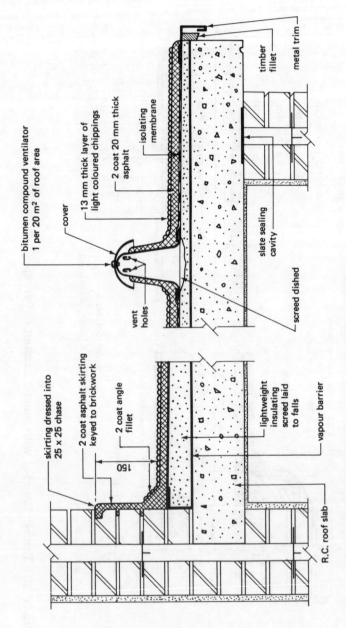

bitumen compound ventilator 1 per 20 m² of roof area

cover

13 mm thick layer of light coloured chippings

2 coat 20 mm thick asphalt

isolating membrane

vent holes

timber fillet

metal trim

slate sealing cavity

screed dished

skirting dressed into 25 x 25 chase

2 coat asphalt skirting keyed to brickwork

2 coat angle fillet

150

lightweight insulating screed laid to falls

vapour barrier

R.C. roof slab

Fig. V.21   R.C. flat roof with mastic asphalt covering

ash, exfoliated vermiculite and expanded perlite. The thickness should not be less than 40 mm and it may be necessary to apply to the screed a 1:4 cement/sand topping to provide the necessary surface finish. Mix ratios of 1:8—10 are generally recommended for screeds of approximately 1 100 kg/m^3 density using foamed slag, sintered pulverised fuel ash and expanded clay, whereas a 1:5 mix is recommended for exfoliated vermiculite and expanded perlite aggregates giving a low density of under 640 kg/m^3. When mixing screeds containing porous aggregates high water/cement ratios are required to give workable mixes and therefore the screeds take a long time to dry out. Since the average rainfall in this country generally exceeds the rate of evaporation it is not always possible to ensure that the screed has dried out before the impermeable finish is applied. This entrapped moisture will tend to vaporise and cause an upward pressure during warm weather which could result in lifting or blistering of the mastic asphalt layer. To overcome this problem roof ventilators can be fixed to help relieve the pressure (see Fig. IV.12).

Moisture in the form of a vapour will tend to rise within the building and condense on the underside of the covering or within the thickness of the insulating material. If an insulating material becomes damp its efficiency will decrease and if composed of organic material it can decompose. To overcome this problem a vapour barrier of a suitable impermeable material such as polythene sheet or aluminium foil should be placed beneath the insulating layer. Care must be taken to see that all laps are complete and sealed and that the edges of the insulating material are similarly protected against the infiltration of moisture vapour.

The surface of a flat roof being fully exposed will gain heat by solar radiation and if insulated will be raised to a temperature in excess of the air or ambient temperature, since the transfer of heat to the inside of the building has been reduced by the insulating layer. The usual method employed to reduce the amount of solar heat gain of the covering is to cover the upper surface of the roof with light coloured chippings to act as a reflective finish. Suitable aggregates are white spar, calcined flint, white limestone or any light coloured granite of 12 mm size and embedded in a bitumen compound.

36
Lead-covered flat roofs

Lead as a building material has been used extensively for over 5 000 years and is obtained mainly from the mineral galena of which Australia, Canada, Mexico and the USA are the main producers. The raw material is mined, refined to a high degree of purity and then cast into bars or pigs which can be used to produce lead sheet, pipe and extruded products.

Lead is a durable and dense material (11 340 kg/m^3) of low strength but is very malleable and can be worked cold into complicated shapes without fracture. In common with other non-ferrous metals lead oxidises on exposure to the atmosphere and forms a thin protective film or coating over its surface. When in contact with other metals there is seldom any corrosion by electrolysis and therefore fixing is usually carried out by using durable copper nails.

For flat roofs milled lead sheet complying with the recommendations of BS 1178 is used. This sheet is supplied in rolls of a standard width of 2.400 m with lengths up to 12.000 m. For easy identification lead sheet carries a colour coding for each code number thus:

BS Code No.	Thickness (mm)	Colour
3	1.25	green
4	1.80	blue
5	2.24	red
6	2.50	black
7	3.15	white
8	3.55	orange

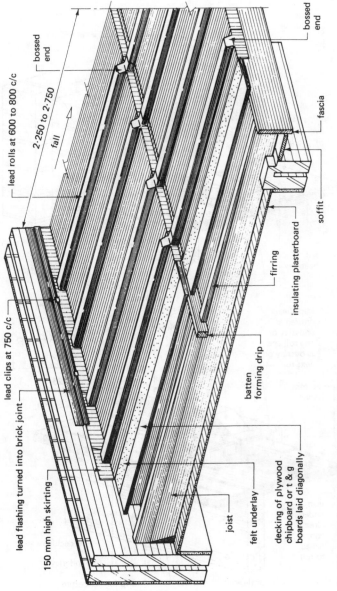

bossed end

bossed end

lead rolls at 600 to 800 c/c

2·250 to 2·750

fall

fascia

soffit

firring

insulating plasterboard

batten forming drip

joist

felt underlay

decking of plywood chipboard or t & g boards laid diagonally

lead clips at 750 c/c

lead flashing turned into brick joint

150 mm high skirting

Fig. V.22 Typical layout of lead flat roof

345

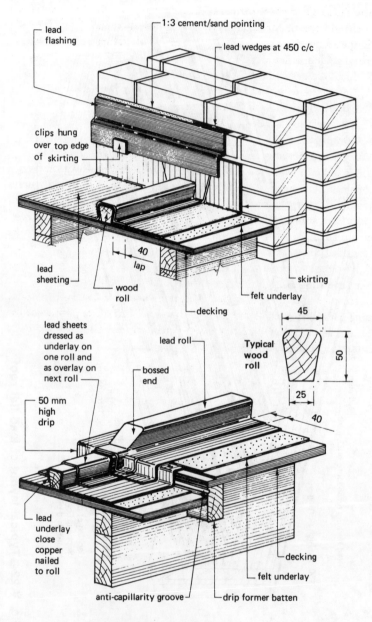

lead flashing

1:3 cement/sand pointing

lead wedges at 450 c/c

clips hung over top edge of skirting

40 lap

lead sheeting

wood roll

decking

felt underlay

skirting

lead sheets dressed as underlay on one roll and as overlay on next roll

lead roll

bossed end

Typical wood roll

45

50

25

40

50 mm high drip

lead underlay close copper nailed to roll

anti-capillarity groove

drip former batten

decking

felt underlay

Fig. V.23   Lead flat roof details

The code number is derived from the former imperial notation of
5 lb/ft^2 = No. 5 lead.

The thickness or code number of lead sheet for any particular situation
will depend upon the protection required against mechanical damage and
the shape required. The following thickness can therefore be considered as
a general guide for flat roofs:

1. Small areas without foot traffic No. 4 or 5.
2. Small areas with foot traffic No. 5, 6 or 7.
3. Large areas with or without foot traffic No. 5, 6 or 7.
4. Flashings No. 4 or 5.
5. Aprons No. 4 or 5.

Milled lead sheet may be used as a covering over timber or similar
deckings and over smooth screeded surfaces. In all cases an underlay of
felt or stout building paper should be used to reduce frictional resistances,
decrease surface irregularities and in the case of a screeded surface isolate
the lead from any free lime present which may cause corrosion. Provision
must also be made for the expansion and contraction of the metal covering.
This can be achieved by limiting the area and/or length of the sheets being
used. It is recommended that the area of any one piece should not exceed
2.25 m^2 and the length should not exceed 2.500 m. Joints which can
accommodate the anticipated thermal movements are in the form of rolls
running parallel to the fall and drips at right angles to the fall positioned so
that they can be cut economically from a standard sheet; for layout and
construction details see Figs. V.22 and V.23.

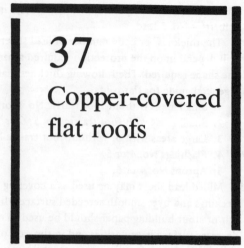

37
Copper-covered flat roofs

Copper, like lead, has been used as a building material for many centuries. It is obtained from ore deposits which have to be mined, crushed and ground to a fine powder. Generally by a system of flotation the copper dust, which is about 4% of the actual rock mined, is separated from the waste materials and transferred to the smelting furnace and then cast into cakes of blister copper or thick slabs called anodes. The metal is now refined and formed into copper wire, strip, sheets and castings. Copper is also used to form alloys which are used in making components for the building industry. Common alloys are copper- and zinc-forming brass and copper- and tin-forming bronze.

Copper is a dense material (8 900 kg/m^3) which is highly ductile and malleable and can be cold worked into the required shape or profile. The metal hardens with cold working but its original dead soft temper can be restored by the application of heat with a blow-lamp or oxy-acetylene torch and quenching with water or by natural air cooling. If the dead soft temper is not maintained the hardened copper will be difficult to work and may fracture. On exposure to the atmosphere copper forms on its upper surface a natural protective film or patina, which varies in colour from green to black, making the copper beneath virtually inert.

For covering flat roofs rolled copper complying with the recommendations of BS 2870 is generally specified. Rolled copper is available in three forms:

Sheet: flat material of exact length, over 0.15 mm up to and including 10.0 mm thick and over 450 mm in width.

Strip: material over 0.15 mm up to and including 10.0 mm thick and any

348

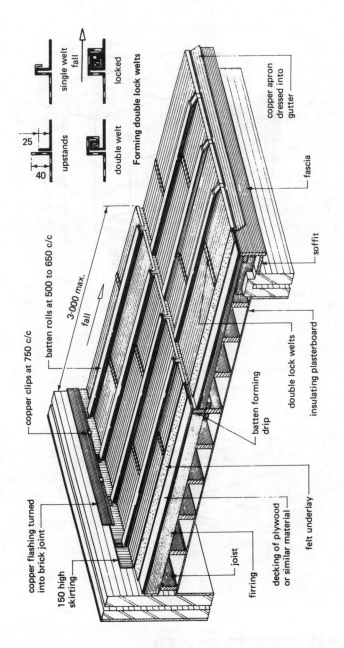

single welt
fall
locked

upstands
25
40
double welt

Forming double lock welts

copper apron dressed into gutter

fascia

soffit

batten rolls at 500 to 650 c/c

3·000 max.

fall

copper clips at 750 c/c

double lock welts

insulating plasterboard

batten forming drip

copper flashing turned into brick joint

150 high skirting

joist

firring

decking of plywood or similar material

felt underlay

Fig. V.24   Typical layout of copper flat roof

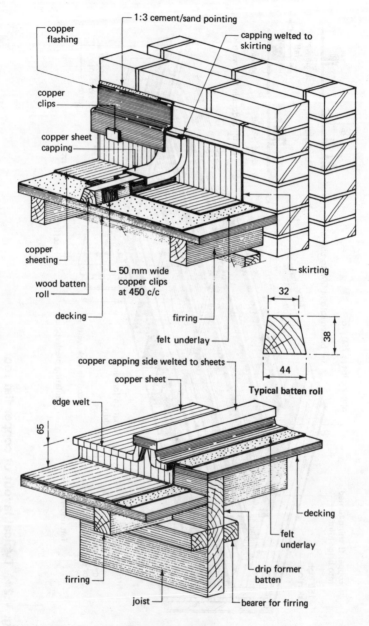

Fig. V.25 Copper flat roof details

width, and generally not cut to length. It is usually supplied in coils but can be obtained flat or folded.

Foil: material 0.15 mm thick and under, of any width supplied flat or in a coil; because of its thickness foil has no practical application in the context of roof coverings.

In general copper strip is used for flashings and damp-proof courses, whereas sheet or strip can be considered for general roof covering application, widths of 600 and 750 mm, according to thickness, are used with a standard length of 1.800 m.

The copper sheet or strip is laid in bays between rolls or standing seams to allow for the expansion of the covering. Standing seams are not recommended for pitches under $5°$ since they may be flattened by foot traffic and become a source of moisture penetration due to a capillary action. The recommended maximum bay widths for the common sheet thicknesses used are:

    0.45 m thick – maximum bay width 600.
    0.60 m thick – maximum bay width 600.
    0.70 m thick – maximum bay width 750.

Transverse joints in the form of double lock welts should be used on all flat roofs. The welts should be sealed with a mastic or linseed oil before being folded together. Cross welts may be staggered or continuous across the roof when used in conjunction with batten rolls (see Fig. V.24).

The substructure supporting the covering needs to be permanent, even, smooth and laid to the correct fall of not less than 1 : 60 (40 mm in 2.400 m). With a concrete structural roof a lightweight screed is normally laid to provide the correct surface and fall; if the screed has a sulphate content it will require sealing coats of bitumen to prevent any moisture present forming dilute acids which may react with the copper covering. The 50 x 25 dovetail battens inserted in the screed to provide the fixing medium for the batten rolls should be impregnated with a suitable wood preservative. Timber flat roofs should be wind tight and free from 'spring' and may be covered with 25 mm nominal tongued and grooved boarding laid with the fall or diagonally, or any other decking material which fulfils the general requirements given above.

On flat roofs drips at least 65 mm high are required at centres not exceeding 3 000 centres to increase the flow of rain water across the roof to the gutter or outlet. To lessen the wear on the copper covering as it expands and contracts a separating layer of felt (BS 747 Type 4A(ii)) should be incorporated into the design. The copper sheet should be secured with copper wire nails not less than 2.6 mm thick and at least 25 mm long; the batten rolls should be secured with well countersunk brass screws. Typical constructional details are shown in Fig. V.25.

38

Factory-type roofs

The general concepts of roof functions, construction techniques and methods of covering have already been covered in the first two years of study (see relevant sections in volumes 1 and 2 Construction Technology). The study in a typical third year course is concerned mainly with certain building types such as small to large industrial or factory buildings. Buildings of this nature set the designer two main problems:

1. Production layout at floor level and the consequent need for large unobstructed areas by omitting as far as practicable internal roof supports such as load bearing walls and columns.
2. Provision of natural daylight from the roof over the floor area below.

The amount of useful daylight which can penetrate into a building from openings in side and end walls is very limited and depends upon such factors as height of windows above the working plane or surface, sizes of windows and the arrangement of windows. Buildings with spans in excess of 18.000 m will generally need some form of overhead supplementary lighting. In single storey buildings this can take the form of glazed rooflights but in multi-storey buildings the floors beneath the top floor will have to have the natural daylighting from the side and end walls supplemented by permanent artificial lighting.

The factors to be considered when designing or choosing a roof type or profile for a factory building in terms of rooflighting are:

1. Amount of daylight required.
2. Spread of daylight required over the working plane.

3. Elimination of solar heat gain and solar glare.

The amount of daylight required within a building is usually based upon the daylight factor, which can be defined as 'the illumination at a specific point indoors expressed as a percentage of the simultaneous horizontal illumination outdoors under an unobstructed overcast sky'. The minimum daylight factor for factories recommended by the 'Illuminating Engineering Society' is 5%. The designer can calculate the daylight factor achieved by a particular roof profile and glass area by using a BRS daylight protractor (see BRE Digests 41 and 42) but an approximation can be made by using a rule of thumb which gives a ratio of one-fifth glass to floor area thus:

assume a daylight factor of 6% is required over a floor area of 500 m^2

then area of glass required

= daylight factor × floor area × 5

= $\frac{6}{100}$ × 500 × 5

= 150 m^2

It must be emphasised that this one-fifth rule of thumb is a preliminary design aid and the result obtained should be checked by a more precise method.

For an even spread of light over the working plane the ratio between the spacing and height of the rooflights is particularly important unless a monitor roof is used which gives a reasonable even spread of light by virtue of its shape. For typical ratios see Fig. V.27. If the ratios shown in Fig. V.27 are not adopted the result could give a marked difference in illumination values at the working plane, resulting in light and darker areas.

Another factor to be considered when planning rooflighting is the amount of obstruction to natural daylighting which could be caused by services and equipment housed in the roof void. Also the gradual deterioration of the effectiveness of the glazed areas due to the collection of dirt on the surfaces.

The elimination of solar heat gain and solar glare can be achieved in a number of ways such as fitting reflective louvres to the glazed areas and treating the surface with a special thin paint wash to act as a diffusion agent. Probably the best solution is to choose a roof profile such as the northlight roof which is orientated away from the southern aspect. It must be appreciated that due to other factors such as site size, site position and/or planning requirements the last solution is not always possible.

NORTHLIGHT ROOFS

This form of roof profile is asymmetric; the north orientated face is pitched at an angle of between 50° and 90° and is covered with glass, usually in the form of patent glazing. The south orientated face of the roof is pitched at an angle of between 20° and 30° and is covered with profiled asbestos cement sheeting or similar lightweight covering attached to purlins. The structural roof members can be of timber, steel or precast concrete formed as a plane frame in single or multi-span format and spaced at 4.500 to 6.000 m centres according to the spanning properties of the purlins – see Fig. V.26.

Single span northlight roofs in excess of 12.000 m span are generally unacceptable since the void formed by the roof triangulation is very large and since it is not divorced from the main building it has to be included in the total volume to be heated. The solution is to use a series of smaller plane frames to form a multiple northlight roof which will reduce the total volume of the roof considerably. It should be noted that there is practically no difference in the total roof area to be covered whichever method is adopted; however, multiple roofs will require more fittings in the form of ridge pieces, eaves closures and gutters.

Using a system of multiple northlight frames a valley is formed between each plane frame and this must be designed to collect and discharge the surface water run off from both sides of the valley – see Fig. V.28. Internal support for the ends of adjacent frames can be obtained by using a valley beam spanning over internal columns, the spacing of these supporting columns being determined by the spanning properties of the chosen beam. If the number of internal columns required becomes unacceptable in terms of floor layout and circulation an alternative arrangement can be used consisting of a lattice girder housed between the apex and bottom tie of the roof frame – see Fig. V.26. This method will enable economic spans of up to 30.000 m to be achieved. The inclusion of a lattice girder in this position will create a cantilever northlight roof truss and if the same principle is adopted for a multispan symmetrically pitched roof truss it is usually called an umbrella roof. The only real disadvantage of this alternative method is a slight increase in shadow casting caused by the lattice beam member and a small increase in long term maintenance such as painting.

MONITOR ROOFS

This form of roof is basically a flat roof with raised portions glazed on two faces which are called monitor lights. The roof covering is usually some form of lightweight metal decking covered with asphalt or built-up roofing felt. The glazed areas, like those in a

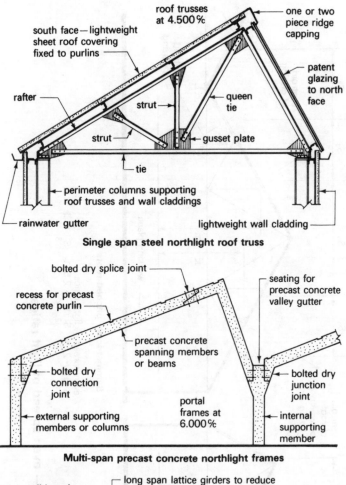

Single span steel northlight roof truss

roof trusses at 4.500%

one or two piece ridge capping

south face—lightweight sheet roof covering fixed to purlins

patent glazing to north face

rafter

strut

queen tie

strut

gusset plate

tie

perimeter columns supporting roof trusses and wall claddings

rainwater gutter

lightweight wall cladding

Multi-span precast concrete northlight frames

bolted dry splice joint

recess for precast concrete purlin

seating for precast concrete valley gutter

precast concrete spanning members or beams

bolted dry connection joint

bolted dry junction joint

external supporting members or columns

portal frames at 6.000%

internal supporting member

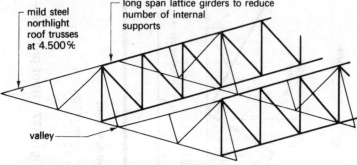

mild steel northlight roof trusses at 4.500%

long span lattice girders to reduce number of internal supports

valley

Fig. V.26   Typical northlight roof profiles

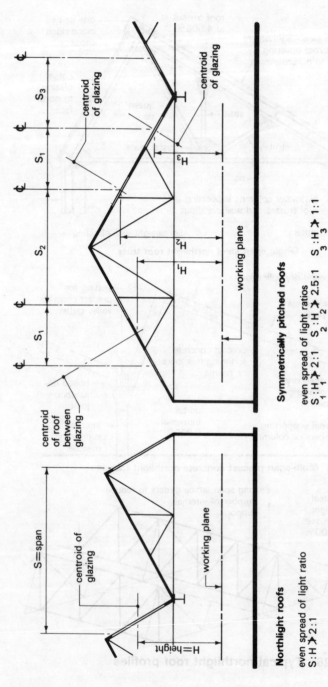

Northlight roofs

even spread of light ratio
$S : H \geq 2 : 1$

S = span

centroid of glazing

centroid of roof between glazing

working plane

H = height

Symmetrically pitched roofs

even spread of light ratios
$S_1 : H_1 \geq 2 : 1$ $S_2 : H_2 \geq 2.5 : 1$ $S_3 : H_3 \geq 1 : 1$

centroid of glazing

centroid of glazing

working plane

Fig. V.27 Roof profiles — even spread of light ratios

356

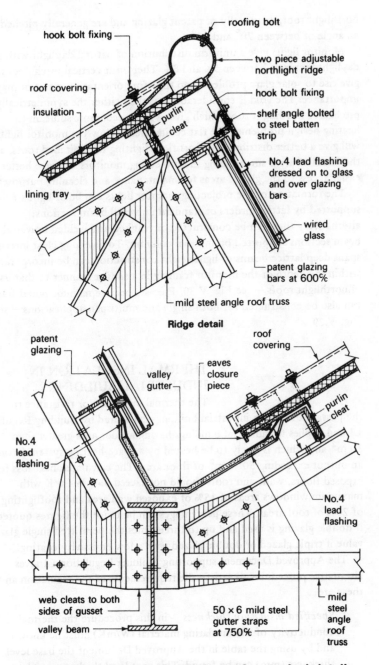

Ridge detail

roofing bolt

hook bolt fixing

two piece adjustable
northlight ridge

roof covering

insulation

hook bolt fixing

shelf angle bolted
to steel batten
strip

purlin
cleat

No.4 lead flashing
dressed on to glass
and over glazing
bars

lining tray

wired
glass

patent glazing
bars at 600%

mild steel angle roof truss

patent
glazing

roof
covering

valley
gutter

eaves
closure
piece

purlin
cleat

No.4
lead
flashing

No.4
lead
flashing

web cleats to both
sides of gusset

valley beam

50 × 6 mild steel
gutter straps
at 750%

mild
steel
angle
roof
truss

Fig. V.28   Multiple northlight roof — typical details

northlight roof, are usually of patent glazing and are generally pitched at an angle of between 70° and 90°.

Monitor lights give a uniform distribution of natural daylight with a daylight factor of between 5% and 8%. Their near vertical pitch does not give rise to solar glare problems and therefore orientation is not of major importance. The void is considerably less than either the symmetrically pitched or northlight roof which gives a more economic solution to heating design problems. The flat ceiling areas below the monitor lights will give a better distribution of artificial lighting than pitched roofs, also the flat roof areas surrounding the projecting monitors will give better and easier access to the glazed areas for maintenance and cleaning purposes.

The formation of the projecting monitor lights can be of light steelwork supported by lattice girders or standard universal beam sections; alternatively they can be constructed of cranked and welded universal beam sections supported on internal columns. To give long clear internal spans deep lattice beams of lightweight construction can be incorporated within the depth of the monitor framing in a similar manner to that used in northlight roofs — see Fig. V.30. Precast concrete monitor portal frames can also be constructed for both single and multispan applications — see Fig. V.29.

THERMAL INSULATION IN INDUSTRIAL BUILDINGS

The thermal insulation or resistance to the passage of heat in industrial buildings is covered by Building Regulation L1 and applies to industrial buildings having a floor area greater than 30 m^2 and which is likely to be heated by a space-heating system having an output exceeding 50 W/m^2 of floor area. The calculated heat loss for exposed floors, walls and roofs should not exceed 0.45 W/m^{2}K with maximum windows areas of 15% of exposed wall area and rooflighting of 20% of roof areas. These percentages can be twice the figures quoted if double glazing is used and up to three times the permitted single glazing value if triple glazed or double glazed with a low emissivity coating.

The Approved Document supporting Building Regulation L1 gives alternative procedures to satisfy the requirements of the regulation and these are:

1. *Specified insulation thickness* — in this procedure the thermal conductivity of the insulating material (W/mK) must be known and by using the table in the Approved Document the base level thickness (mm) can be found. This base level thickness can be reduced by taking into account features of construction such as

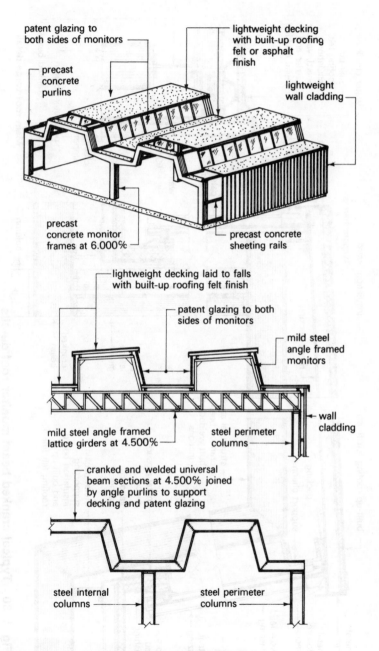

Fig. V.29 Typical monitor roof profiles

359

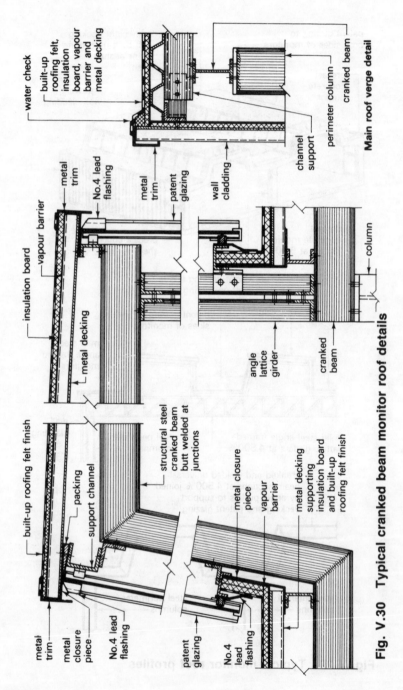

water check

built-up roofing felt, insulation board, vapour barrier and metal decking

perimeter column

cranked beam

channel support

wall cladding

patent glazing

metal trim

Main roof verge detail

metal trim

vapour barrier

insulation board

metal trim

No.4 lead flashing

metal decking

column

angle lattice girder

cranked beam

built-up roofing felt finish

metal trim

metal closure piece

packing

support channel

No.4 lead flashing

structural steel cranked beam butt welded at junctions

metal closure piece

vapour barrier

No.4 lead flashing

metal decking supporting insulation board and built-up roofing felt finish

patent glazing

Fig. V.30 Typical cranked beam monitor roof details

air space and plaster finishes. The reductions allowed are quoted in the same table as the insulation base level thickness.

2. *Calculated trade off* — this procedure is designed to overcome the problem of limiting window and rooflight percentages and maximum 'U' values for walls, roofs and exposed floors. The allowable heat loss (W/K) is calculated for the proposed building using the limitations set out in Building Regulation L1 and this is compared with the calculated rate of heat loss for the actual proposed design. If it can be shown that the rate of heat loss in the proposed design is less than the allowable rate of heat loss the proposal would satisfy the regulation requirements.

3. *Calculated energy use* — this procedure can be used for buildings other than dwellings and allows a full trade-off between glazed and solid areas taking into account any useful heat gains such as solar heat gains, artificial lighting and industrial processes. The calculations must prove that the annual energy consumption taking into account heat gains would be no greater than if the values given in Building Regulation L1 had been used.

The heat loss from within a building is affected by the temperature difference between the internal and external environments, and to comply with various Acts and to create good working conditions, it is desirable to maintain a minimum temperature for various types of activities.

The ideal working temperature for any particular task is a subjective measure but as a guide the following internal temperatures recommended by the Institution of Heating and Ventilating Engineers are worth considering:

1. Sedentary work   — 18.4°C minimum
2. Light work   — 15.6°C minimum
3. Heavy work   — 12.8°C minimum

The Factories Act 1961 requires a minimum temperature, after the first hour, of 15.6°C to be maintained except where sedentary work takes place where a higher temperature would be required.

The advantages which can be gained by having a well insulated roof are:

1. Lower fuel bills.
2. Reduced capital outlay on heating equipment.
3. Better working conditions for employees and hence better working relationships.

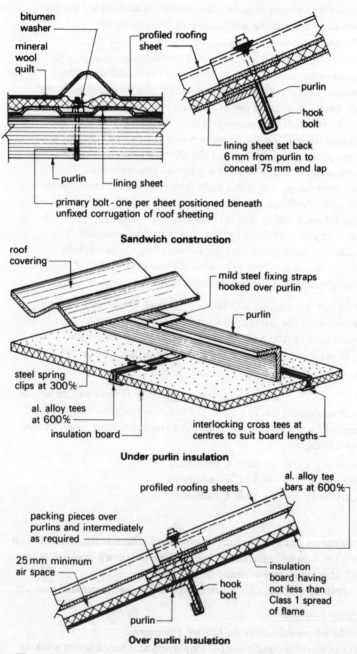

Sandwich construction

Under purlin insulation

Over purlin insulation

Fig. V.31 Factory roofs — typical insulation details

The initial cost of a building will be higher if a good standard of thermal insulation is specified and installed but in the long term an overall saving is usually experienced, the increased capital outlay being recovered within the first five or so years by the savings made in running costs.

The inclusion of certain materials within a factory roof to comply with the thermal insulation requirements of this Act may introduce into the structure a fire hazard. To this end the Act stipulates that the exposed surfaces of insulation materials used, even if within a cavity must be at least class 1 spread of flame as defined in BS 476 : Part 7. By using suitable materials or combinations of materials the risk of fire and fire spread in factories can be reduced considerably (Fig. V.31). The main objective, in common with all fires in buildings, is to contain the fire to the vicinity of the outbreak should a fire occur.

Factories can have compartment type walls with automatic closing fire-resistant doors, but if open planning is required other precautions will be necessary. The roof volume can be divided into cells by fitting permanent fire barriers within the triangulated profile of the roof structure with non-combustible materials such as fibre cement sheet suitably fire stopped within the profile of the roof covering. Beneath these fire barriers can be rolled curtains of fire fibre cloth controlled by a fusible link or similar fire detection device which will allow the curtain to fall forming a fire screen from roof to floor in the advent of a fire. A similar curtain could also be positioned in the longitudinal direction under a valley beam.

Using the above method a fire can be contained for a reasonable period within the confined area but it can also create another problem or hazard, that of smoke logging. The smoke generated by a fire will rise to the roof level and then start to circulate within the screened compartment, completely filling the volume of the confined section within a short space of time which apart from the hazard to people trying to escape can present the fire fighters with the following problems:

1. Difficulty in breathing.
2. Prevention from seeing the source of the fire.
3. Detecting nature of outbreak.
4. Assessing extent of outbreak.

A method of overcoming this problem is to have automatic high level ventilators which will allow the smoke to escape rapidly and thus give the fire fighters a chance to see clearly and enable them to deal with the outbreak. The use of ventilators, to overcome the problems of smoke logging, will of course introduce more air which aids combustion but

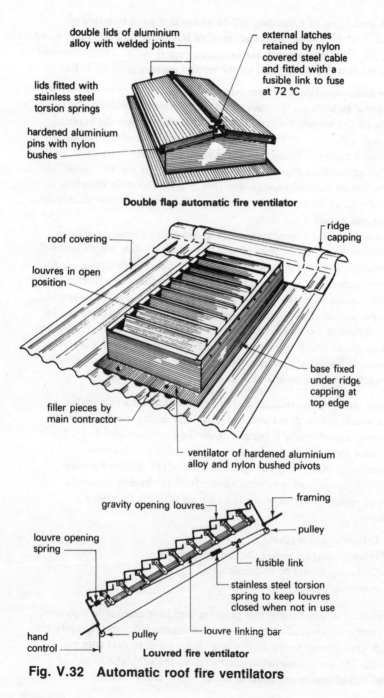

double lids of aluminium alloy with welded joints

external latches retained by nylon covered steel cable and fitted with a fusible link to fuse at 72 °C

lids fitted with stainless steel torsion springs

hardened aluminium pins with nylon bushes

Double flap automatic fire ventilator

ridge capping

roof covering

louvres in open position

base fixed under ridge capping at top edge

filler pieces by main contractor

ventilator of hardened aluminium alloy and nylon bushed pivots

gravity opening louvres

framing

louvre opening spring

pulley

fusible link

stainless steel torsion spring to keep louvres closed when not in use

hand control

pulley

louvre linking bar

Louvred fire ventilator

Fig. V.32 Automatic roof fire ventilators

this does not have the same negative effect as smoke logging since the volume of air in this type of building is usually so vast that the introduction of more air will have very little effect on the intensity of the outbreak.

The design and position of automatic roof ventilators is normally the prerogative of a specialist designer but as a guide the total area of opening ventilators should be between 0.5% and 5% of the floor area depending on the likely area of fire. The essential requirements for an automatic ventilator are:

1. It must open in the event of a fire, a common specification being when the heat around the ventilator reaches a temperature of 68°C.
2. Weatherproof under normal circumstances.
3. Easy to fix and blend with chosen roof covering material and profile.

Many automatic fire ventilators are designed to act as manually controlled ventilators under normal conditions — typical examples are shown in Fig. V.32.

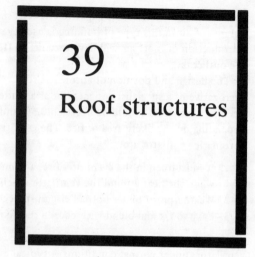

39
Roof structures

Roof design and construction, like foundations, is an extensive topic and it is therefore the usual practice to build up a student's comprehension by including in each year of study one particular aspect of roofing techniques. In this text the roofs under consideration are those suitable for large clear spans using structural timber, steelwork and reinforced concrete. Other types of roof construction are described in earlier volumes.

LARGE SPAN TIMBER ROOFS
A wide variety of timber roofs are available for both medium and large spans and these can be classified under three headings:

1. Pitched trusses.
2. Flat top girders.
3. Bowstring trusses.

Pitched trusses: these are two-dimensional triangulated designed frames spaced at 4.500 to 6.000 centres with spans up to 30.000. The pitch should have a depth to span ratio of $1:5$ or steeper and be chosen to suit roof coverings. The basic construction follows that of an ordinary domestic small span roof truss as shown in Fig. V.5 except that the arrangement and number of struts and ties will vary according to the type of truss being used — see Fig. V.33.

Flat top girders: basically these are lattice beams of low pitch spaced at

4.500 to 6.000 centres and can be economically used for spans up to 45.000 with a depth to span ratio of 1 : 8 to 1 : 10. Construction details are similar to those of the roof truss in that the joints and connections are usually made with timber connectors and bolts. The main advantage of this form of roof structure is the reduction in volume of the building which should result in savings in the heating installation required and in the running costs. Typical flat top girder outlines are shown in Fig. V.33 and typical construction details are shown in Fig. V.34.

Bowstring trusses: these trusses are basically a lattice girder with a curved upper chord and are spaced at 4.500 to 6.000 centres with an economic span range of up to 75.000. The depth-to-span ratio is usually 1 : 6 to 1 : 8 with the top chord radius approximately equal to the span. They can be constructed from solid segmental timber pieces but the chords are usually formed from laminated timber with solid timber struts and ties forming the lattice members — see Fig. V.35 for typical details. The older form of bowstring truss known as a belfast truss which has interlacing struts and ties is not very often specified because it is difficult to analyse fully the stresses involved and although relatively small section timbers can be used they are very expensive in labour costs.

Choice of truss and timber

To decide upon the most suitable and economic truss to be specified for any given situation the following should be considered:

1. Availability of suitable timber in the sizes required.
2. Cost of alternative timber.
3. Design and fabrication costs.
4. Transportation problems and costs.
5. On-site assembly and erection problems and costs.
6. Roof covering material availability and costs.
7. Architectural design considerations.

It may be possible that the consideration given to the last item may well outweigh some of the economic solutions found for the preceding items.

The timber specified should comply with the recommendations of BS 5268 and any subsequent amendments. The specified timber must also satisfy the structural stability requirements of Building Regulation A1 and the recommendations set out in the supportive Approved Document A. This means that unless a designer, manufacturer or builder makes special arrangements to obtain approval, each piece of timber used in a strength application in any form of building will have to be stress graded to the

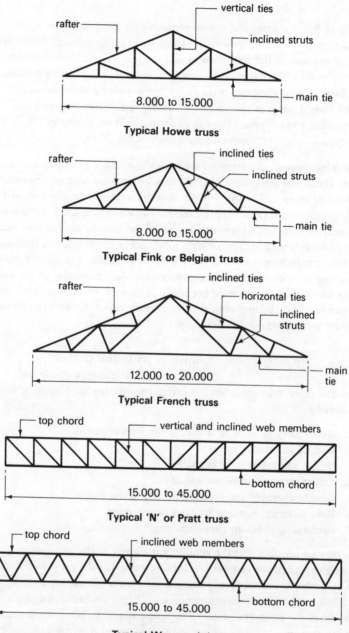

Fig. V.33 Typical large span truss and girder types

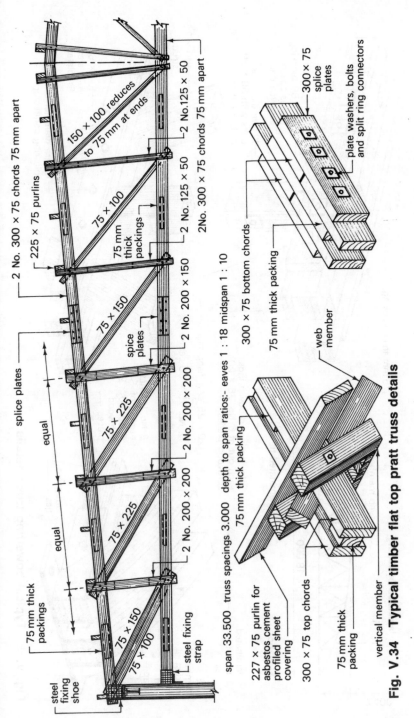

2 No. 300 × 75 chords 75 mm apart

150 × 100 reduces to 75 mm at ends

225 × 75 purlins

2 No.125 × 50

75 × 100

2 No. 125 × 50

75 mm thick packings

2 No. 300 × 75 chords 75 mm apart

splice plates

75 × 150

2 No. 200 × 150

splice plates

2 No. 200 × 200

75 × 225

equal

2 No. 200 × 200

75 × 225

equal

75 mm thick packings

75 × 150

steel fixing shoe

75 × 100

steel fixing strap

span 33.500 truss spacings 3.000 depth to span ratios:- eaves 1 : 18 midspan 1 : 10

300 × 75 splice plates

plate washers, bolts and split ring connectors

300 × 75 bottom chords

75 × 75 thick packing

web member

75 mm thick packing

227 × 75 purlin for asbestos cement profiled sheet covering

300 × 75 top chords

75 mm thick packing

vertical member

Fig. V.34 Typical timber flat top pratt truss details

369

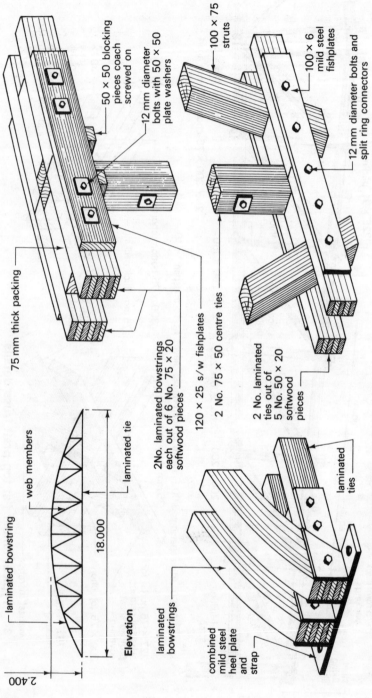

50 × 50 blocking pieces coach screwed on

12 mm diameter bolts with 50 × 50 plate washers

100 × 75 struts

100 × 6 mild steel fishplates

12 mm diameter bolts and split ring connectors

75 mm thick packing

2 No. laminated bowstrings each out of 6 No. 75 × 20 softwood pieces

120 × 25 s/w fishplates

2 No. 75 × 50 centre ties

2 No. laminated ties out of 5 No. 50 × 20 softwood pieces

laminated bowstring

web members

laminated tie

laminated tie

18.000

2.400

Elevation

laminated bowstrings

laminated ties

combined mild steel heel plate and strap

Fig. V.35   Typical bowstring truss details

370

grades set out in BS 5268. Reference should also be made to BS 4978 'Timber Grades for Structural Use'. This standard was published after collaboration between engineers, scientists and the timber trades of the United Kingdom and the main timber-exporting countries such as Canada, Finland and Sweden. The stress grading can be carried out either visually or by machine and the permissible stresses for the various grades and species are set out in BS 5268 : Part 2.

Visual stress grading is carried out by the knot area ratio (KAR) method in which the proportion of cross-section occupied by the projected area of knots is assessed. Any pieces of timber where the KAR is less than one-fifth is graded special structural (SS) and for pieces of timber where the KAR is between one-fifth and one-half is graded as general structural (GS) or SS depending on whether a margin condition exists. Any piece of timber with a KAR exceeding one-half is automatically rejected as being suitable for structural work.

Machine stress grading relies on the correlation between the strength of timber and its stiffness. These grading machines work on one of two principles, those which apply a fixed load and measure the deflection and those which measure the load required to cause a fixed deflection. The gradings obtained are designated MGS (machine general structural) or MSS (machine special structural) and these grades are comparable to the visual grades given above. BS 5268 also describes two further grades, namely M50 and M75 which means that these pieces of timber have been graded as having 50 or 75% of the strength of a clear specimen of a similar species.

All graded timber must be marked so that it can be immediately identified by the specifier, supplier and user. Visually graded timber is marked at least once within the length of each piece with GS or SS together with a mark to indicate the company or grader responsible for the grading. Machine stress graded timber should be marked MGS, MSS, M50 or M75 at least once within the length of each piece together with a mark to indicate species, grading machine used, BSI Kitemark and the relevant BS number, namely BS 4978. Machine stress graded timber can also be colour coded with a coloured dye at one end or a series of dashes throughout the length. The colour coding used being:

Green	MGS
Purple	MSS
Blue	M50
Red	M75

The most readily available species of structural softwoods are imported redwood, imported whitewood and imported commercial western hemlock. Other suitable structural softwoods are only generally available in small quantities. Structural softwoods are supplied in standard lengths commencing at 1.800 and increasing by 300 mm increments to a maximum length of 6.300 with section size within the range of the table given in BS 4471. When specifying stress graded timber the following points should be considered:

1. Species.
2. Section size.
3. Length.
4. Preparation requirements.
5. Stress grade.
6. Moisture content.

Jointing
Connections between structural timber members may be made by:

1. Nails.
2. Screws.
3. Glue and nails or screws.
4. Truss plates.
5. Bolts.
6. Bolts and timber connectors.

Nails and screws are usually used in conjunction with plywood gussets and like the truss or gangnail plates are usually confined to the small to medium span trusses.

The usual fixings such as nails, screws and bolts have their own limitations. Cut nails will generally have a greater holding power than round wire nails due to the higher friction set up by their rough sharp edges and also the smaller disturbance of the timber grain. The joint efficiency of nails may be as low as 15% owing to the difficulty in driving sufficient numbers of them within a given area to obtain the required shear value. Screws have a greater holding power than nails but are dearer in both labour and material costs.

Joints made with a rigid bar such as a bolt usually have low efficiency due to the low shear strength of timber parallel to the grain and the unequal distribution of bearing stress along the shank of the bolt. The weakest point in these connections is where the high stresses are set up around the bolt and various methods have been devised to overcome this problem. The solution lies in the use of timber connectors which are designed to increase the bearing area of the bolts as described below.

Toothed plate timber connector — sometimes called bulldog connectors.

These are used to form an efficient joint without the need for special equipment and are suitable for all types of connections, especially when small sections are being used. To form the connection the timber members are held in position and drilled for the bolt to provide a bolt hole with 1.5 mm clearance. If the timber is not too dense the toothed connectors can be embedded by tightening up the permanent bolts. In dense timbers or where more than three connectors are used the embedding pressure is provided by a high tensile steel rod threaded at both ends. Once the connectors have been embedded the rod is removed and replaced by the permanent bolt.

Split ring timber connectors — these connectors are suitable for any type of structure, timbers of all densities, are very efficient and develop a high strength joint. The split ring is a circular band of steel with a split tongue and groove vertical joint. A special boring and cutting tool is required to form the bolt hole and the grooves in the face of the timber into which the connector is inserted making the ring independent of the bolt itself. The split in the ring ensures that a tight fit is achieved on the timber core but at the same time being sufficiently flexible to give a full bearing on the timber outside the ring when under heavy load.

Shear plate timber connector — these are counterparts of the split ring connectors, they are housed flush into the timber members and are used for demountable structures. See Fig. V.36 for typical timber connector details.

LARGE SPAN STEEL ROOFS

The roof types given for large span timber roofs can also be designed and fabricated using standard structural steel sections. Span ranges and the spacings of the frames or lattice girders are similar to those given for timber roofs. Connections can be of traditional gusset plates to which the struts and ties would be bolted or welded; alternatively an all-welded construction is possible especially if steel tubes are used to form the struts and ties. Large span steel roofs can also take the form of space decks and space frames.

Space decks

A space deck is a structural roofing system designed to give large clear spans with wide column spacings. It is based on a simple repetitive unit consisting of an inverted pyramid frame which can be joined to similar frames to give spans of up to 22.000 for single spanning designs and up to 33.000 for two-way spanning roofs. These basic

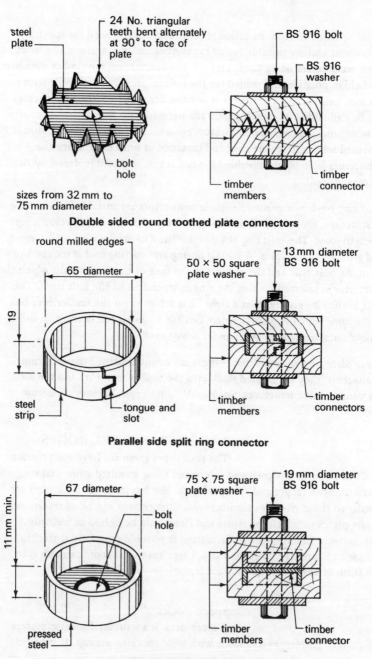

steel plate

24 No. triangular teeth bent alternately at 90° to face of plate

BS 916 bolt

BS 916 washer

bolt hole

timber members

timber connector

sizes from 32 mm to 75 mm diameter

Double sided round toothed plate connectors

round milled edges

50 × 50 square plate washer

13 mm diameter BS 916 bolt

65 diameter

19

steel strip

tongue and slot

timber members

timber connectors

Parallel side split ring connector

75 × 75 square plate washer

19 mm diameter BS 916 bolt

67 diameter

11 mm min.

bolt hole

pressed steel

timber members

timber connector

Shear plate connectors

Fig. V.36 Typical BS 1579 timber connectors

units are joined together at the upper surface by bolting adjacent angle framing together and by fixing threaded tie bars between the apex couplers — see Fig. V.37.

Edge treatments include vertical fascia, mansard and cantilever. Rooflights can be fixed directly over the 1.200 x 1.200 modular upper framing — see Fig. V.38. The roof can be laid or screeded to falls, or alternatively a camber to form the falls can be induced by tightening up the main tie bars. The space deck roof can be supported by steel or concrete columns or fixed to padstones or ring beams situated at the top of load-bearing perimeter walls.

The most usual and economic roof covering for space decks is an insulated decking of wood wool slabs 50 mm thick covered with three layers of built-up roofing felt with a layer of reflective chippings. The void created by the space deck structure can be used to house all forms of services and the underside can be left exposed or covered in with an attached or suspended ceiling. The units are usually supplied with a basic protective paint coating applied by a dipping process after the units have been degreased, shot blasted and phosphate treated to provide the necessary key.

The simplicity of the space deck unit format eliminates many of the handling and transportation problems encountered with other forms of large span roof. A complete roof can usually be transported on one lorry by stacking the pyramid units one inside the other. Site assembly and erection is usually carried out by a specialist sub-contractor who must have access to the whole floor area. Assembly is rapid with a small labour force which assembles the units as beams in the inverted position, turns them over and connects the whole structure together by adding the secondary tie bars. Two general methods of assembly and erection can be used:

1. Deck is assembled on the completed floor immediately below the final position and lifted directly into its final position.
2. Deck is assembled outside the perimeter of the building and lifted in small sections to be connected together in the final position — this is generally more expensive than method 1.

The main contractor has to provide prior to assembly and erection a clear, level and hard surface to the perimeter of the proposed building as well as to the whole floor area to be covered by the roof. These surfaces must be capable of accepting the load from a 25 tonne mobile crane. The main contractor is also responsible for unloading, checking, storing and protecting the units during and after erection and for providing all necessary temporary works and plant such as scaffolds, ladders and hoists. The site procedures and main contractor responsibilities set out above in

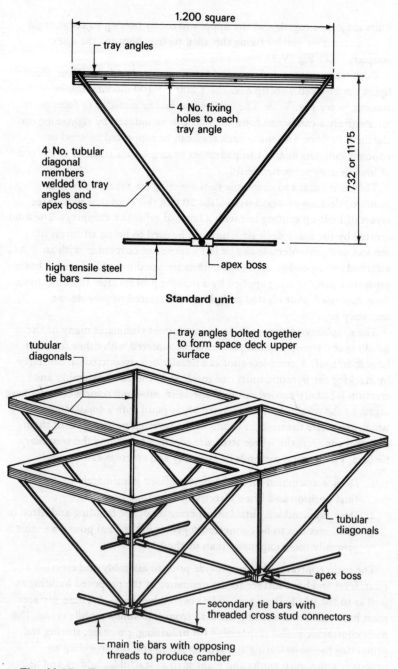

Fig. V.37 Typical 'Space Deck' standard units

Standard unit

1.200 square

tray angles

4 No. fixing holes to each tray angle

4 No. tubular diagonal members welded to tray angles and apex boss

high tensile steel tie bars

apex boss

732 or 1175

tray angles bolted together to form space deck upper surface

tubular diagonals

tubular diagonals

apex boss

secondary tie bars with threaded cross stud connectors

main tie bars with opposing threads to produce camber

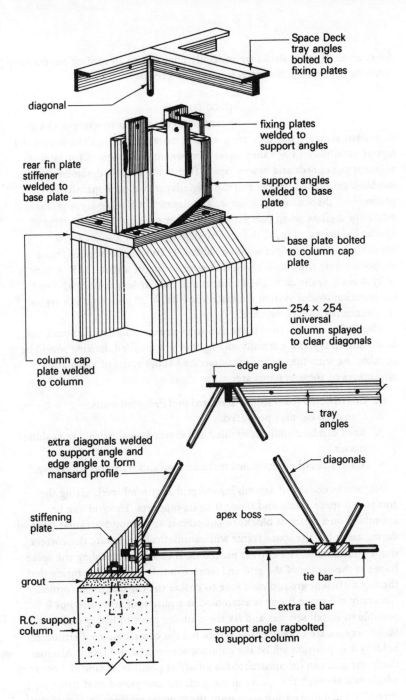

Space Deck
tray angles
bolted to
fixing plates

diagonal

fixing plates
welded to
support angles

rear fin plate
stiffener
welded to
base plate

support angles
welded to base
plate

base plate bolted
to column cap
plate

254 × 254
universal
column splayed
to clear diagonals

column cap
plate welded
to column

edge angle

tray
angles

extra diagonals welded
to support angle and
edge angle to form
mansard profile

diagonals

stiffening
plate

apex boss

grout

extra tie bar

R.C. support
column

support angle ragbolted
to support column

Fig. V.38   Typical 'Space Deck' edge fixing details

the context of space decks are generally common to all roofing contractors involving specialist sub-contractors and materials.

Space frames

Space frames are similar to space decks in their basic concept but they are generally more flexible in their design and layout possibilities since their main component is the connector joining together the chords and braces. Space frames are usually designed as a double-layer grid as opposed to the single-layer grid used mainly for geometrical shapes such as a dome. The depth of a double-layer grid is relatively shallow when compared with other structural roof systems of similar loadings and span, the span-to-depth ratio for a space frame supported on all its edges would be about 1 : 20 whereas a space frame supported near the corners would require a ratio of about 1 : 15. A variety of systems is available to the architect and builder and the British Steel Corporation Nodus system illustrated in Figs. V.39 and V.40 is a typical representative of such systems.

The claddings used in conjunction with a space frame roof should not be unduly heavy and normally any lightweight profiled decking would be suitable. As with the space decks described above some of the main advantages of these systems are:

1. Constructed from simple standard prefabricated units.
2. Units can be mass produced.
3. Roof can be rapidly assembled and erected on site using semi-skilled labour.
4. Small sizes of components make storage and transportation easy.

Site works consist of assembling the grid at ground level, lifting the completed space frame and fixing it to its supports. The grid can be assembled on a series of blocks to counteract any ground irregularities and during assembly the space frame will automatically generate the correct shape and camber. The correct procedure is to start assembling the space frame at the centre of the grid and work towards the edges ensuring that there is sufficient ground clearance to enable the camber to be formed. Generally the space frame is assembled as a pure roof structure but it is possible to install services and fix the cladding prior to lifting and fixing. Mobile cranes are usually employed to lift the completed roof structure holding it in position whilst the columns are erected and fixed. Alternatively the grid can be constructed in an offset position around the columns which pass through the spaces in the grid, the completed roof structure is then lifted and moved sideways onto the support seatings on top of the columns.

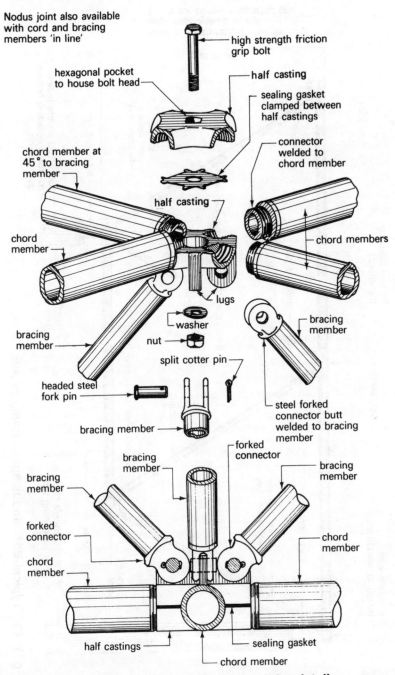

Nodus joint also available with cord and bracing members 'in line'

high strength friction grip bolt

hexagonal pocket to house bolt head

half casting

sealing gasket clamped between half castings

connector welded to chord member

chord member at 45° to bracing member

half casting

chord member

chord members

lugs

washer

nut

split cotter pin

bracing member

headed steel fork pin

bracing member

steel forked connector butt welded to bracing member

forked connector

bracing member

bracing member

bracing member

forked connector

chord member

chord member

half castings

sealing gasket

chord member

Fig. V.39 Typical BSC Nodus System joint details.

379

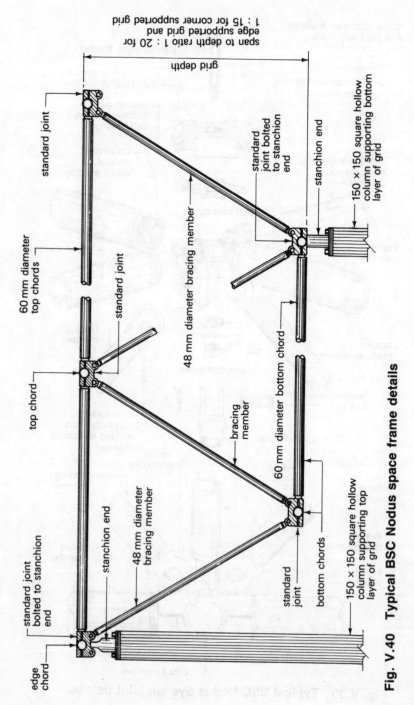

Fig. V.40 Typical BSC Nodus space frame details

span to depth ratio 1 : 20 for
edge supported grid and
1 : 15 for corner supported grid

grid depth

standard joint

standard
joint bolted
to stanchion
end

stanchion end

150 × 150 square hollow
column supporting bottom
layer of grid

60 mm diameter
top chords

standard joint

48 mm diameter bracing member

top chord

bracing
member

60 mm diameter bottom chord

48 mm diameter
bracing member

stanchion end

standard
joint

bottom chords

150 × 150 square hollow
column supporting top
layer of grid

standard joint bolted to stanchion
end

edge
chord

380

SHELL ROOFS

A shell roof may be defined as a structural curved skin covering a given plan shape and area, the main points being:

1. Primarily a structural element.
2. Basic strength of any particular shell is inherent in its shape.
3. Quantity of material required to cover a given plan shape and area is generally less than other forms of roofing.

The basic materials which can be used in the formation of a shell roof are concrete, timber and steel. Concrete shell roofs consist of a thin curved reinforced membrane cast *in situ* over timber formwork whereas timber shells are usually formed from carefully designed laminated timber and steel shells are generally formed using a single layer grid. Concrete shell roofs although popular are very often costly to construct since the formwork required is usually purpose made from timber and which is in itself a shell roof and has little chance of being re-used to enable the cost of the formwork to be apportioned over several contracts.

A wide variety of shell roof shapes and types can be designed and constructed but they can be classified under three headings:

1. Domes.
2. Vaults.
3. Saddle shapes and conoids.

Domes: in their simplest form these consist of a half sphere but domes based on the ellipse, parabola and hyperbola are also possible. Domes have been constructed by architects and builders over the centuries using individually shaped wedge blocks or traditional timber roof construction techniques. It is therefore the method of construction together with the materials employed rather than the geometrical setting out which has changed over the years.

Domes are double curvature shells which can be rotational and are formed by a curved line rotating around a vertical axis or they can be translational domes which are formed by a curved line moving over another curved line — see Figs. V.41 and V.42. Pendentive domes are formed by inscribing within the base circle a polygon and cutting vertical planes through the true hemispherical dome.

Any dome shell roof will tend to flatten due to the loadings and this tendency must be resisted by stiffening beams or similar to all the cut edges. As a general guide domes which rise in excess of one-sixth of their diameter will require a ring beam. Timber domes like their steel counterparts are usually constructed on a single-layer grid system and covered with a suitable thin skin membrane.

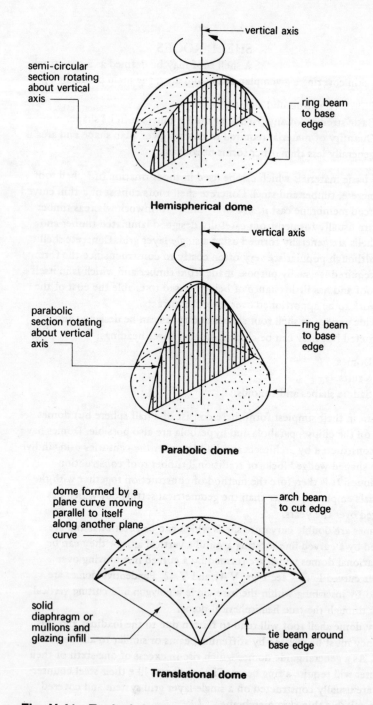

Hemispherical dome

vertical axis

semi-circular
section rotating
about vertical
axis

ring beam
to base
edge

Parabolic dome

vertical axis

parabolic
section rotating
about vertical
axis

ring beam
to base
edge

Translational dome

dome formed by a
plane curve moving
parallel to itself
along another plane
curve

arch beam
to cut edge

solid
diaphragm or
mullions and
glazing infill

tie beam around
base edge

Fig. V.41 Typical dome roof shapes 1

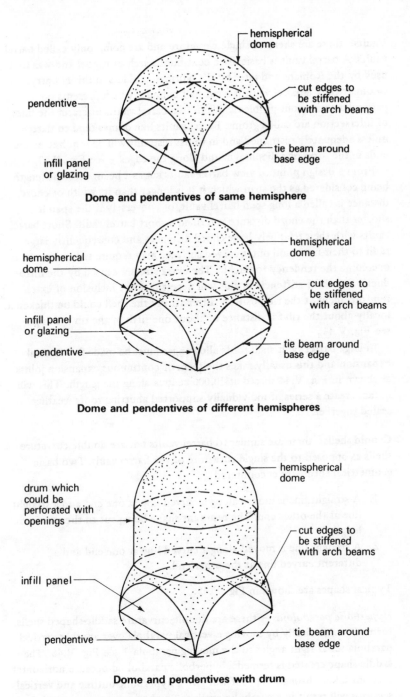

Fig. V.42　Typical dome roof shapes 2

383

Vaults: these are shells of single curvature and are commonly called barrel vaults. A barrel vault is basically a continuous arch or tunnel and was first used by the Romans and later by the Norman builders in this country. Geometrically a barrel vault is a cut half cylinder which presents no particular setting out problems. When two barrel vaults intersect the lines of intersection are called groins. Barrel vaults like domes tend to flatten unless adequately restrained and in vaults restraint will be required at the ends in the form of a diaphragm and along the edges — see Fig. V.43.

From a design point of view barrel vaults act as a beam with the length being considered as the span which if it is longer than its width or chord distance is called a long span barrel vault, or conversely if the span is shorter than the chord distance is termed a short barrel vault. Short barrel vaults with their relatively large chord distances and consequently large radii to their inner and outer curved surfaces may require stiffening ribs to overcome the tendency to buckle. The extra stresses caused by the introduction of these stiffeners or ribs will necessitate the inclusion of extra reinforcement at the rib position, alternatively the shell could be thickened locally about the rib for distance of about one-fifth of the rib spacing — see Fig. V.43.

In large barrel vault shell roofs allowances must be made for thermal expansion and this usually takes the form of continuous expansion joints as shown in Fig. V.44 spaced at 30.000 centres along the length. This will in fact create a series of individually supported abutting roofs weather sealed together.

Conoid shells: these are similar to barrel vaults but are double curvature shells as opposed to the single curvature of the barrel vault. Two basic geometrical forms are encountered:

1. A straight line is moved along a curved line at one end and a straight line at the other end. The resultant shape being cut to the required length.
2. A straight line is moved along a curved line at one end and a different curved line at the other end.

Typical shapes are shown in Fig. V.45.

Hyperbolic paraboloids — these are double-curvature saddle-shaped shells formed geometrically by moving a vertical parabola over another vertical parabola set at right angles to the moving parabola — see Fig. V.46. The saddle shape created is termed a hyperbolic paraboloid because horizontal sections taken through the roof will give a hyperbolic outline and vertical sections will result in a parabolic outline. To obtain a more practical shape

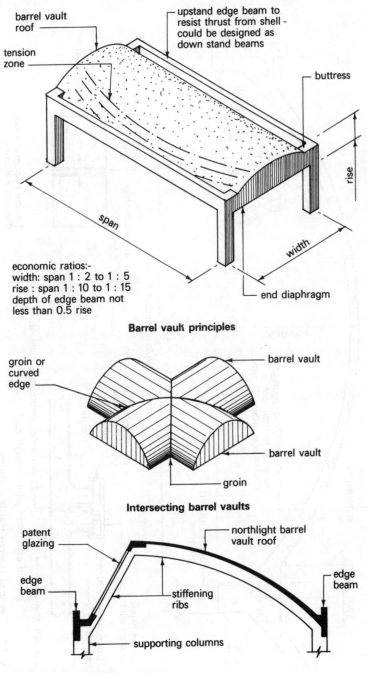

barrel vault roof

tension zone

upstand edge beam to resist thrust from shell - could be designed as down stand beams

buttress

rise

span

width

economic ratios:-
width: span 1 : 2 to 1 : 5
rise : span 1 : 10 to 1 : 15
depth of edge beam not
less than 0.5 rise

end diaphragm

Barrel vault principles

groin or curved edge

barrel vault

barrel vault

groin

Intersecting barrel vaults

patent glazing

northlight barrel vault roof

edge beam

stiffening ribs

edge beam

supporting columns

Fig. V.43   Typical barrel vaults

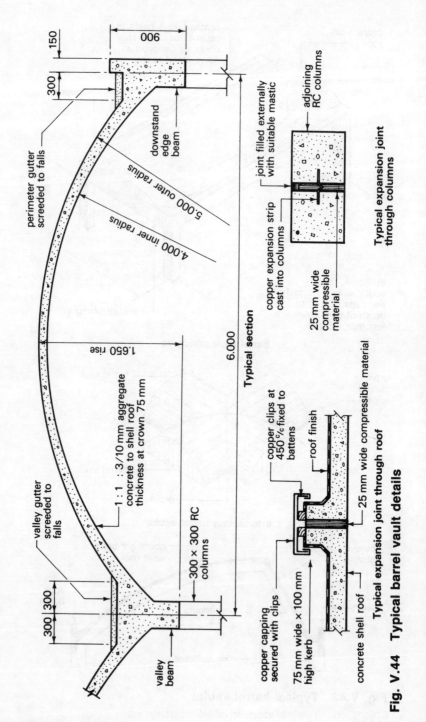

Typical section

150

006

300

perimeter gutter screeded to falls

downstand edge beam

5.000 outer radius

4.000 inner radius

1.650 rise

6.000

1 : 1 : 3/10 mm aggregate concrete to shell roof thickness at crown 75 mm

valley gutter screeded to falls

300 × 300 RC columns

300 | 300

valley beam

Typical expansion joint through columns

adjoining RC columns

joint filled externally with suitable mastic

copper expansion strip cast into columns

25 mm wide compressible material

Typical expansion joint through roof

copper clips at 450 % fixed to battens

roof finish

25 mm wide compressible material

copper capping secured with clips

75 mm wide × 100 mm high kerb

concrete shell roof

Fig. V.44 Typical barrel vault details

386

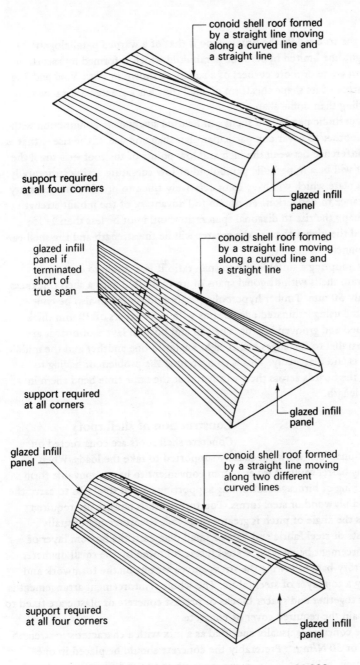

conoid shell roof formed by a straight line moving along a curved line and a straight line

support required at all four corners

glazed infill panel

glazed infill panel if terminated short of true span

conoid shell roof formed by a straight line moving along a curved line and a straight line

support required at all corners

glazed infill panel

glazed infill panel

conoid shell roof formed by a straight line moving along two different curved lines

support required at all four corners

glazed infill panel

typical widths for all types between 12.000 and 30.000 spans very often made less than width

Fig. V.45 Typical conoid shell roof types

than the true saddle the usual shape is that of a warped parallelogram or straight line limited hyperbolic paraboloid which is formed by raising or lowering one or more corners of a square as shown in Figs. V.46 and V.47. By virtue of its shape this form of shell roof has a greater resistance to buckling than dome shapes.

Hyperbolic paraboloid shells can be used singly or in conjunction with one another to cover any particular plan shape or size. If the rise — that is the difference between the high and low points of the roof — is small the result will be a hyperbolic paraboloid of low curvature acting structurally like a plate which will have to be relatively thick to provide the necessary resistance to deflection. To obtain full advantage of the inbuilt strength of the shape the rise to diagonal span ratio should not be less than 1 : 15; indeed the higher the rise the greater will be the strength and the shell can be thinner.

By adopting a suitable rise-to-span ratio it is possible to construct concrete shells with diagonal spans of up to 35.000 with a shell thickness of only 50 mm. Timber hyperbolic paraboloid roofs can also be constructed using laminated edge beams with three layers of 20 mm thick tongued and grooved boards. The top and bottom layers of boards are laid parallel to the edges but at right angles to one another and the middle layer is laid diagonally. This is to overcome the problem of having to twist the boards across their width and at the same time bend them in their length.

Construction of shell roofs

Concrete shell roofs are constructed on traditional formwork adequately supported to take the loads. When casting barrel vaults it is very often convenient to have a movable form consisting of birdcage scaffolding supporting curved steel ribs to carry the curved plywood or steel forms. Top formwork is not usually required unless the angle of pitch is greater than 45°. Reinforcement usually consists of steel fabric and bars of small diameter, the bottom layer of reinforcement being welded steel fabric followed by the small diameter trajectory bars following the stress curves set out on the formwork and finally a top layer of steel fabric. The whole reinforcement arrangement is wired together and spacer blocks of precast concrete or plastic are fixed to maintain the required cover of concrete.

The concrete is usually specified as a mix with a characteristic strength of 25 or 30 N/mm^2. Preferably the concrete should be placed in one operation in 1 m wide strips commencing at one end and running from edge beam to edge beam over the crown of the roof. A wet mix should be placed around the reinforcement followed by a floated drier mix. Thermal

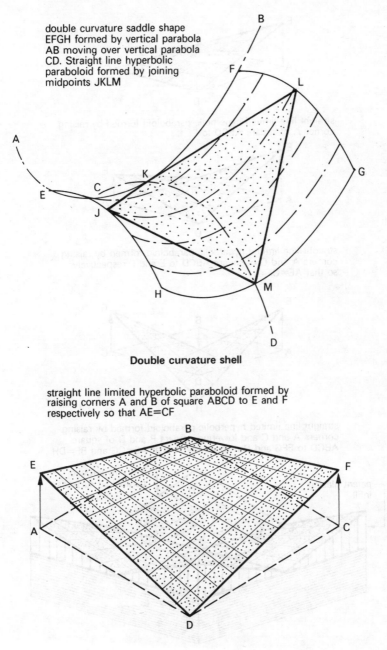

double curvature saddle shape
EFGH formed by vertical parabola
AB moving over vertical parabola
CD. Straight line hyperbolic
paraboloid formed by joining
midpoints JKLM

Double curvature shell

straight line limited hyperbolic paraboloid formed by
raising corners A and B of square ABCD to E and F
respectively so that AE=CF

Hyperbolic paraboloid shell

Fig. V.46 Hyperbolic paraboloid roof principles

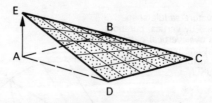

straight line limited hyperbolic paraboloid formed by raising corner A of square ABCD to E

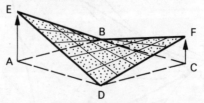

straight line limited hyperbolic paraboloid formed by raising corners A and C of square ABCD to E and F respectively so that AE≠CF

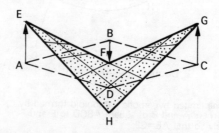

straight line limited hyperbolic paraboloid formed by raising corners A and C and lowering corners B and D of square ABCD to EFG and H respectively so that AE=CG and BF=DH

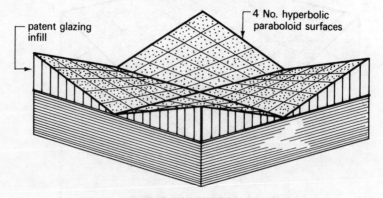

hyperbolic paraboloids combined to form single roof

Fig. V.47 Typical hyperbolic paraboloid roof shapes

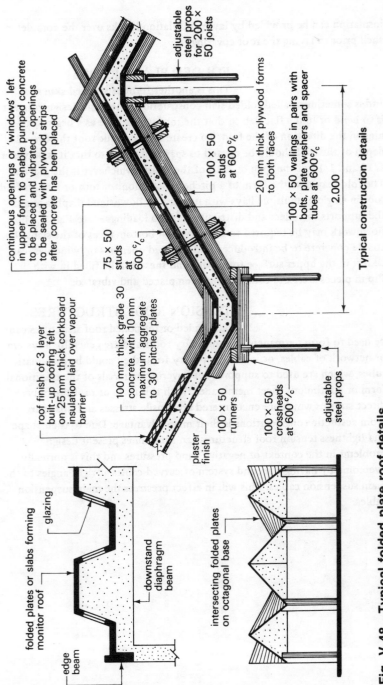

Fig. V.48 Typical folded plate roof details

Typical construction details

adjustable steel props for 200 × 50 joists

continuous openings or 'windows' left in upper form to enable pumped concrete to be placed and vibrated - openings to be sealed with plywood strips after concrete has been placed

100 × 50 studs at 600 °/c

20 mm thick plywood forms to both faces

100 × 50 walings in pairs with bolts, plate washers and spacer tubes at 600 °/c

75 × 50 studs at 600 °/c

2.000

roof finish of 3 layer built-up roofing felt on 25 mm thick corkboard insulation laid over vapour barrier

100 mm thick grade 30 concrete with 10 mm maximum aggregate to 30° pitched plates

100 × 50 runners

100 × 50 crossheads at 600 °/c

adjustable steel props

plaster finish

folded plates or slabs forming monitor roof

glazing

downstand diaphragm beam

edge beam

intersecting folded plates on octagonal base

insulation can be provided by laying insulation blocks over the completed shell prior to laying the roof covering.

FOLDED PLATE ROOFS

This is another form of stressed skin roof and is sometimes called folded slab construction. The basic design concept is to bend or fold a flat slab so that the roof will behave as a beam spanning in the direction of the fold. To create an economic roof the overall depth of the roof should be related to span and width so that it is between 1/10 and 1/15 of the span or 1/10 of the width, whichever is the greater. The fold may take the form of a pitched roof, monitor roof or a multi-fold roof in single or multiple bays with upstand or downstand diaphragms at the supports to collect and distribute the slab loadings – see Fig. V.48. Formwork may be required to both top and bottom faces of the slabs. To enable concrete to be introduced and vibrated openings or 'windows' can be left in the upper surface formwork and these will be filled in with slip-in pieces after the concrete has been placed and vibrated.

TENSION ROOF STRUCTURES

Suspended or tensioned roof structures can be used to form permanent or temporary roofs and are generally a system or network of cables, or in the temporary form they could be pneumatic tubes, which are used to support roof covering materials of the traditional form or continuous sheet membranes. With this form of roof the only direct stresses which are encountered are tensile stresses and this apart from aesthetic considerations is their main advantage. Due to their shape and lightness tension roof structures can sometimes present design problems in the context of negative wind pressures and this is normally overcome by having a second system of curved cables at right angles to the main suspension cables. This will in effect prestress the main suspension cables.

Bibliography

Relevant BS — British Standards Institution.
Relevant CP — British Standards Institution.
Building Regulations — HMSO.
Relevant BRE Digests — HMSO.
Relevant Advisory Leaflets — DOE.
DOE Construction Issues 1—17 — DOE.
R. Barry. *The Construction of Buildings*. Crosby Lockwood & Sons Ltd.
Mitchells Building Construction Series. B. T. Batsford Ltd.
W. B. McKay. *Building Construction*, Vols. 1 to 4. Longman.
Specification. The Architectural Press.
A. J. Elder. *A. J. Guide to the Building Regulations*. The Architectural Press.
Construction Safety. The National Federation of Building Trades Employers.
R. Llewelyn Davies and D. J. Petty. *Building Elements*. The Architectural Press.
Cecil C. Handisyde. *Building Materials*. The Architectural Press.
W. Fisher Cassie and J. H. Napper. *Structure in Building*. The Architectural Press.
Handbook on Structural Steelwork. The British Constructional Steelwork Association Ltd. and the Constructional Steel Research and Development Organisation.
L. V. Leech. *Structural Steelwork for Students*. Butterworths.
B. Boughton. *Reinforced Concrete Detailers Manual*. Crosby Lockwood & Sons Ltd.

A Guide to Scaffolding Construction and Use. Scaffolding (Great Britain) Ltd.

The Plumbers Handbook — Lead Development Association.

Copper Roofing — Copper Development Association.

Application of Mastic Asphalt — Mastic Asphalt Council and Employers Federation.

Relevant A. J. Handbooks. The Architectural Press.

Relevant manufacturers' catalogues contained in the Barbour Index and Building Products Index Libraries.

Index